Cities and Cinema

Films about cities abound. They provide fantasies for those who recognize their city and those for whom the city is a faraway dream or nightmare. How does cinema rework city planners' hopes and city dwellers' fears of modern urbanism? Can an analysis of city films answer some of the questions posed in urban studies? What kinds of vision for the future and images of the past do city films offer? What are the changes that city films have undergone?

Cities and Cinema puts urban theory and cinema studies in dialogue. The book's first section analyzes three important genres of city films that follow in historical sequence, each associated with a particular city, moving from the city film of the Weimar Republic to the film noir associated with Los Angeles and the image of Paris in the cinema of the French New Wave. The second section discusses socio-historical themes of urban studies, beginning with the relationship of film industries and individual cities, continuing with the portrayal of war-torn and divided cities, and ending with the cinematic expression of utopia and dystopia in urban science fiction. The last section negotiates the question of identity and place in a global world, moving from the portrayal of ghettos and barrios to the city as a setting for gay and lesbian desire, to end with the representation of the global city in transnational cinematic practices.

The book suggests that modernity links urbanism and cinema. It accounts for the significant changes that city film has undergone through processes of globalization, during which the city has developed from an icon in national cinema to a privileged site for transnational cinematic practices. It is a key text for students and researchers of Film Studies, Urban Studies, and Cultural Studies.

Barbara Mennel is an Assistant Professor of German Studies and Cinema Studies in the Department of Germanic and Slavic Studies and in the Film and Media Studies Program in the English Department at the University of Florida, Gainesville. She is author of *The Representation of Masochism and Queer Desire in Film and Literature* (2007).

Routledge critical introductions to urbanism and the city

The series is designed to allow undergraduate readers to make sense of, and find a critical way into, urbanism. It will:

- cover a broad range of themes
- introduce key ideas and sources
- allow the author to articulate her/his own position
- introduce complex arguments clearly and accessibly
- bridge disciplines, and theory and practice
- be affordable and well designed.

The series covers social, political, economic, cultural and spatial concerns. It will appeal to students in architecture, cultural studies, geography, popular culture, sociology, urban studies, urban planning. It will be trans-disciplinary. Firmly situated in the present, it also introduces material from the cities of modernity and post-modernity.

Published:

Cities and Consumption – Mark Jayne
Cities and Cultures – Malcolm Miles
Cities and Nature – Lisa Benton-Short and John Rennie Short
Cities and Economies – Yeong-Hyun Kim and John Rennie Short
Cities and Cinema – Barbara Mennel

Forthcoming:

Cities, Politics and Power – Simon Parker
Children, Youth and the City – Kathrin Hörshelmann and Lorraine van Blerk
Cities and Gender – Helen Jarvis, Jonathan Cloke and Paula Kantor

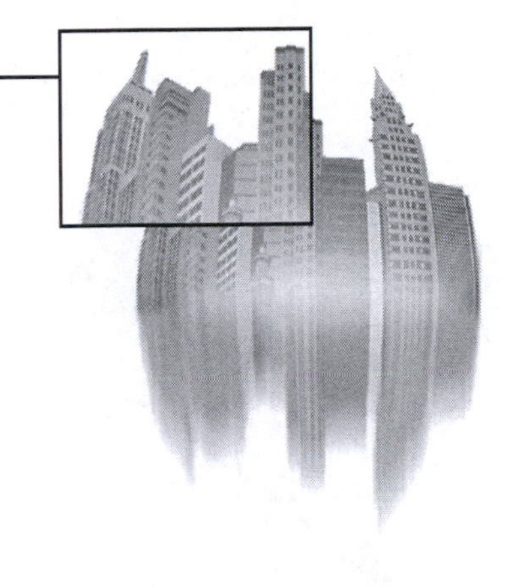

Cities and Cinema

By Barbara Mennel

LONDON AND NEW YORK

First published 2008 by Routledge
2 Park Square, Milton Park, Abingdon, Oxon, OX14 4RN
Simultaneously published in the USA and Canada
by Routledge
270 Madison Avenue, New York, NY 10016

Routledge is an imprint of the Taylor & Francis Group, an informa business

Typeset in Times New Roman by Keystroke,
28 High Street, Tettenhall, Wolverhampton

British Library Cataloguing in Publication Data
A catalogue record for this book is available from the British Library

Library of Congress Cataloging in Publication Data
Mennel, Barbara Caroline.
Cities and cinema / by Barbara Mennel.
p. cm.
Includes bibliographical references and index.
1. Cities and towns in motion pictures. 2. City and town life in motion pictures.
I. Title.
PN1995.9.C513M46 2008
791.43'621732—dc22
2007037747

ISBN10: 0–415–36445–0 (hbk)
ISBN10: 0–415–36446–9 (pbk)
ISBN10: 0–203–01560–6 (ebk)

ISBN13: 978–0–415–36445–4 (hbk)
ISBN13: 978–0–415–36446–1 (pbk)
ISBN13: 978–0–203–01560–5 (ebk)

Contents

Figures

Acknowledgments

First and foremost I have to thank John Rennie Short with whom I had the pleasure of teaching a course on "Cinema and the City" for the Humanities Scholars Program at the University of Maryland, Baltimore County, out of which this volume emerged. I not only learned from his lectures but also cherished our regular meals accompanied by animated conversations about our respective fields. And I am deeply indebted to Amy Abugo Ongiri, without whom this book could not have been written. Long ago she introduced me to Blaxploitation and Hong Kong action film, and during the writing of the book she generously shared her extensive collection of materials with me. She also read drafts of a couple of chapters in a pinch – thanks. More importantly, however, this book is deeply influenced by her thinking about film and popular culture. Ingeborg Majer O'Sickey provided a lake-view in upstate New York for a joint writing–retreat and was a passionate and insightful sounding-board for the project. Her intention to turn the book into a course provided me with an imaginary audience. I am also indebted to Jeffrey Schneider and Jaimey Fisher who both took time out of their busy schedules to read and make incisive comments on sections of the manuscript that left their imprints on concepts integral to the book's argument.

During the fall of 2006 I was nourished and sustained by the students in two courses I was privileged to teach at the University of Florida, Gainesville, a graduate course, "Theories of Globalization and the Cinema," and an undergraduate course, "Literature, Film, and the Arts of Berlin." Graduate students Hendrik Aulbach, Jonathan Barnes, Heather Bigley, Jo Carlisle, Matthew Feltman, Claudia Hoffmann, Dominik Jaschke, Yun Jo, James Liner, Peter Mersch, Fayola Neely, James Phillips, Kay Sender, and Rabia Shah offered original and thought-provoking approaches to some of the films discussed in this volume, pushed my thinking on theories of globalization and, most importantly, created community around our shared intellectual pursuit. Equally sharp and committed, undergraduates Arace Assadoghli, Daryl Baginski, Rachael Counce, Benjamin Dorvel, Leah Greenblum, Heather Harr, Delia Hernandez, Jonathan Hill, Joshua

McClellan, Natalie Prager, Chelsea Rhodes, Margot Salinardi, Lila Stone, Haneke van de Esschert, Sarah Wichterman, and Steven Wylie engaged deeply with the culture of Berlin, sharing contagious excitement accompanied by thorough research and analysis.

Among my colleagues at the University of Florida, Susan Hegeman provided substantive feedback and references for early drafts of the first three chapters during a one-semester leave generously provided by my then chairs, Dragan Kujundzic and John Leavey. Andrew M. Gordon kindly discussed science fiction and urban space with me and pointed me in the direction of seminal films and literature, while Sylvie Blum-Reid read and commented on Chapter 3 on Paris with map in hand. Ewa Wampuszyc and Holly Raynard read Chapter 5 and offered important comparative perspectives on war and the city in relationship to Warsaw and Prague respectively, part of their on-going witty comparative European intellectual exchange. Interim chair and historian Jeffrey S. Adler balanced the universe for one academic year when much of the writing took place, and through his sheer presence reminded me of the importance of history (and dogs).

Research in the Film Archive of the Deutsche Kinemathek in Berlin, the British Film Institute in London, and the Bibliothèque du Film in Paris was supported by a Humanities Enhancement Grant and a Research Travel Grant from the College of Liberal Arts and Sciences at the University of Florida and the Department of Germanic and Slavic Studies, under chair Will Hasty. I thank the Photo Archive of the Deutsche Kinemathek, especially Peter Latta, and the British Film Institute for permissions to reprint the stills in this volume.

Working with the partially anonymous reviewers was the kind of positive academic experience I wish on any scholarly writer. Their feedback was, while sometimes critical, always deeply engaging, constructive, and pushed me beyond the horizon of the already written text. I particularly appreciated David B. Clarke – whose work I respect deeply – for lending his name to the project from the outset. I was equally impressed by the reviewer who remained anonymous and who claimed to be "blunt to be helpful," but who was also generous, smart, funny, conscientious, and knowledgeable. Time is one of the things we never have enough of, and to give time to work that appears under the name of others represents an act of particular intellectual generosity. While I may not have fulfilled all their expectations, I hope to have approximated them. Mary Fahnestock-Thomas has helped me navigate writing in English with kindness, generosity, and force. At Routledge, the dream-team of Andrew Mould, Jennifer Page, and Daniel Wadsworth made everything possible.

This book is dedicated to my sister Susan Ursula Mennel. It took me many years to understand that if I am the city mouse, she is the country mouse, even though the deep mud on her boots could have tipped me off much earlier.

Introduction: The founding myth of cinema, or the "train effect"

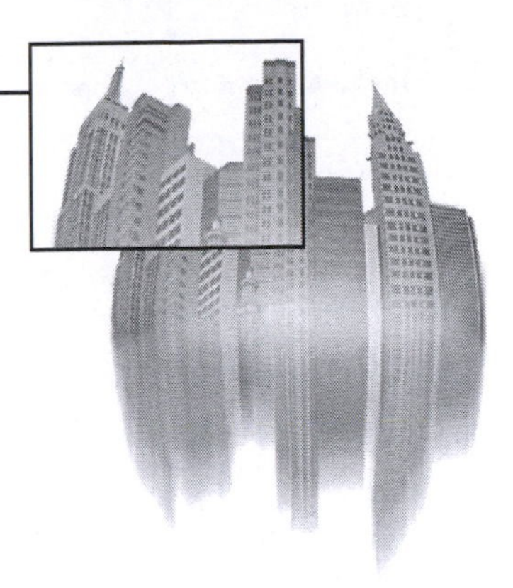

> The cinematograph reigns in the city, reigns over the earth . . . More than the preachings of wise men, the cinematograph has demonstrated to everyone what reality is.
>
> Andrei Bely (1908)

Learning objectives

- Comprehend the early history of cinema
- Conceptualize the role of cities in that history
- Grasp the terms modernity and postmodernity, and national cinema and transnational cinematic practices
- Understand approaches to analyzing cities and films

Paris is the site of the often-reproduced founding myth of cinema: "On December 28, 1895, cinema begins in the basement of the Grand Café, Boulevard des Capucines, Paris," proclaims Vicky Lebeau (1). She refers, of course, to the mythical first public demonstration of the *Cinématographe* by the brothers Lumière who dazzled their audience by projecting moving pictures onto a screen. The city is integral to this story of how cinema began. Lebeau records that at the time, journalists described the experience as "excitement bordering on terror," and on occasion, she concludes, "the terror became panic" (1). According to Lebeau, this was particularly the case at the showing of the Lumières' 50 seconds long, silent short film *The Arrival of a Train at La Ciotat station* (1895), which "is supposed to have had spectators rearing away from the screen, the dread of colliding with the rush of that enormous machine too much for those who succumbed to the hallucination of the image." By conjoining icons of modernity – urbanity, speed, cinema, and the city – in one seminal moment, the often-cited myth reproduces the story that cinema tells of itself: when the lights go off, an illusion appears and seems so real that we forget we are watching moving pictures.

Yuri Tsivian labeled the reaction of panic to an approaching train on the early screen the "train effect" (Bottomore 178). Scholars have demonstrated, however, that the portrayal of entire audiences panicking in terror from seemingly approaching trains exaggerated exceptional individual occurrences of such reactions (see Bottomore; Christie; Clarke and Doel). Reflected in cartoons, literature, and self-reflexively in film itself, the play on representation and reality associated with celluloid train rides had already become a cliché at the turn of the century. Nicholas Hiley believes that the idea of the panicking audience arose in the 1920s and 1930s, two decades after such stories began to circulate in public (Bottomore 184). These narrative revisions serve to inscribe the later audience as more sophisticated readers of the new medium of film.

Stephen Bottomore has concluded from historical film programs that short films depicting train rides were considered more spectacular than other short films of the period and crowned the end of early film showings "as a kind of sensation" (179). Theater-owners exploited and sensationalized extreme physical and emotional responses. At Tony Pastor's theatre in New York an ambulance was on hand for the showing of James H. White's one-minute *The Black Diamond Express* (1896), which was accompanied by train sound effects, after it was reported that two female audience members had "screamed and fainted" at an earlier showing – though it later turned out they had only "nearly fainted" (Bottomore 181). Clearly, viewers at the time had to learn to negotiate the new medium cognitively, to find a balance between believing and not believing in its realness, which is the precondition for the pleasure of watching film. The many references to panic and terror that circulated in the print media, both in serious articles and in advertisements, also indicate the beginning of advertising and its reliance on sensationalism and thrill. Scholars therefore mine the founding myth of cinema for what it says about modernity, which includes changing perceptions of time and space and the creation of a modern audience coded as urban and sophisticated. By the time the famed 1895 screening at the Grand Café took place, the history of film was already under way.

Early film history

While Russian artist Andrei Bely celebrates the Lumière brothers' invention of the cinematograph as unprecedented and radically world-changing, Luis Lumière himself believed that it was "an invention without future" (Christie 95). David B. Clarke and Marcus A. Doel posit that it was only "editing" that saved film from making a brief appearance at the turn of the century and being subsequently forgotten (52). Scholars emphasize the doubts of the early pioneers in film regarding the medium they had invented and advanced in order to counter the "dangers of imputing a teleology to cinema," which would imply looking back at the history of film from our vantage point and presuming a linear development from its inception to the prevalence of visual culture in contemporary society (42).

The early history of cinema is more complex and contradictory than its founding myth suggests and cannot be reduced to a singular moment, a linear development, or even a single place, such as the city of Paris. Audiences had long enjoyed the projection of images onto the screen at private gatherings and public fairs for entertainment and education, for example by means of the magic lantern, which was invented in the seventeenth century and lasted throughout the nineteenth century, until photography was integrated into its use (Monaco 73). The diorama was also a still and flat projection, but lighting and a translucent canvas made it possible to change the picture, for example from day to night.

There were other presentations of moving images that captured audiences. Throughout the nineteenth century, the mechanical organization of still photographs in different pre-filmic cinematic attractions created the illusion of movement. The zoetrope, for example, evoked the perception of motion when photos of consecutive movements were pasted inside a wheel and spun around. The panorama, which surrounded the spectator with projected images, developed into the padorama, the moving panorama. For example, in 1834 a padorama enabled spectators seated in carriages to visually enjoy parts of the Manchester–Liverpool railway, experiencing the pleasure of the simulated train ride long before film was invented. Clarke and Doel believe that by the end of the 1880s "animated photography was not only widely anticipated, but effectively accomplished" (51).

The invention and consumption of still and moving images was accompanied by an interest in the technological reproduction of sound. Thomas Edison invented the kinetograph to accompany the phonograph he had constructed in 1877, one year after Alexander Graham Bell invented the telephone. Yet it was not until 1927 that sound-film was invented. Before that, films were accompanied by a pianist who improvised a score according to different themes, such as a chase or a romantic scene. In the grand film palaces that were built in cities in the period 1910–30, designers created space for an orchestra, and films were accompanied by an original score.

In 1895 the Lumière brothers patented the cinematograph, which importantly combined camera and projector, and demonstrated it to professional colleagues prior to the aforementioned public screening in the Grand Café. As Bottomore points out, "the cinema in these times was often seen as something bordering on the magical" (179). These turn-of-the-century films, which were very short by today's standards, were shown in amusement parks and at traveling variety shows in combination with magic-lantern projections of still pictures or other pre-filmic attractions like the zoetrope or the kinetoscope. They were shot with a static camera and were not edited. They captured moving objects and created entertaining vignettes, endowing dignitaries and current events with historic importance.

The often-repeated story of the "train effect" does justice neither to early audiences, nor to the creativity and inventiveness of the film pioneers and the diversity of early film. The very first short pieces by the Lumière brothers were meant to demonstrate the new medium of film and showed innocuous slices of reality that demonstrated movement. Their titles reflect their documentary nature: *Exiting the Factory* (1895), *Arrival of a Train at La Ciotat Station*, and *Launching of a Boat* (1900). Not until a few years later did films set out to capture more dramatic movement. For example, *Explosion of a Motor Car* (1900) and *How It Feels to Be Run Over* (1900), by Britain's Cecil Hepworth, and *The Paris–Monte Carlo Run in Two Hours* (1905), by France's Georges Méliès were short narratives about motor vehicles involved in races, accidents, and explosions. At the same time there was a move from realism to the fantastic, as in Robert William Paul's animated film *The ? Motorist* (1906), in England, in which a car runs over a policeman, then up the side of a building, and finally takes off into space (see Christie 22). These developments reflected film's ability to depict speed and movement and captured the concurrent phenomenon of traffic, which made it necessary to adjust one's behavior and cognitive reactions in the city. Jeffrey S. Adler describes how in the city of Chicago the number of cars and fatal automobile accidents rose exponentially in the early twentieth century, making the streets of the city dangerous places for drivers and pedestrians alike, and leading the *Chicago Tribune* to warn people of the threat of "auto slaughter" (205; see also 205–17). This new sense of danger was cinematically reworked in comedic and dramatic form in the many short films about automobiles as killing machines.

While many early films about cars, trains, and other moving objects reflected a modern theme, others were more closely related to existing literary genres such as travelogues, comedies, and literary adaptations. Paul's *A Tour through Spain and Portugal* (n.d.), *Come Along, Do!* (1898), and *The Last Days of Pompeii* (1897) are representative examples respectively (see Christie 24). Early animated films such as Paul's *The Haunted Curiosity Shop* (1901) captured the magical possibilities of film, while in America the "'visual newspaper' style" developed, as in Edwin S. Porter's 1901 films *Kansas Saloon Smashers*, about women prohibitionists, and *Terrible Teddy, The Grizzly King*, about Roosevelt (Christie 30). In

Figure 1 F. W. Murnau. *The Last Laugh* (1924): Modern traffic

Italy, Luigi Maggi's *The Count of Montecristo* (1908) was an example of a fiction narrative (Christie 42). This remarkable international and thematic diversity was paradoxically enabled by the lack of established conventions and economic structures. Scholars of the genre film – "familiar stories with familiar characters in familiar situations" – see its beginnings during the same period (Grant xv). They emphasize the American Western, beginning with Porter's *The Great Train Robbery* (1903), and the gangster film, beginning with D. W. Griffith's *The Musketeers of Pig Alley* (1912) (Grant xv–xvi).

Turn-of-the-century film emerged out of the dynamics of two fields: popular entertainment and technological invention. The Lumière brothers were sons of a successful French photographic manufacturer; Robert William Paul was a maker of scientific instruments in England; and Oskar Messter was the son of an optical manufacturer in Germany. Cecil Hepworth, on the other hand, came from the entertainment business, notably from magic-lantern shows, and toured with a mixed slide-and-film presentation before he created a film laboratory and a studio in 1896. Charles Pathé was a "traveling showman" before coming to control "nearly a quarter of the world's film trade" with the French company Pathé Frères (Christie

94). Ferdinand Zecca, the director of Pathé Frères, came from the "Paris 'singing café'" tradition, while the Russian Evgeni Bauer was a graduate of the Moscow Art College (Christie 37). Alice Guy Blanché entered the industry on yet a third path, joining the French company Gaumont as a secretary before she began to direct and supervise the production of films.

During this phase there was no professional differentiation between director, producer, projector, and distributor. Paul, for instance, was exhibitor, supplier, and producer; the Lumière brothers acted as directors, producers, and distributors; and Blanché was secretary and production supervisor, and later founded a production company in the United States (Christie 24). American Charles Urban worked as an international distributor and film producer, but in the early 1920s directed science fiction (Christie 103). Even though figures from this period later became known as specialists, it is important to remember that they often did not start out as such. Even D.W. Griffith, though known primarily for directing films most of his life, began as an actor and writer (Christie 125).

Despite these important innovations, cinema did not follow a straight path to success. Many of the early film pioneers dropped out or failed after roughly a decade of forcefully and successfully advancing the new medium. The Lumière company stopped production in 1903, Edison left the film business in 1918, Paul returned to instrument-making in 1919, Blanché stopped working as a director after returning from emigration to the United States in 1922, and Hepworth was declared bankrupt in 1924 (Christie 67, 24, 78, and 29).

Early film in cities and cities in early film

Contrary to the founding myth of cinema, Paris was not the only city important in the development of film around the turn of the century. Artistic and technological exchange also took place between London, Berlin, Moscow, and New York, and all of them nourished the early development of film. Thus, the growth of cinema was intimately tied to the growth of cities, and the cities were also associated with the development of movie theaters as urban sites of entertainment and distraction. Films alternated with live performances in music halls and vaudeville theaters, and there were "touring film shows" called "peep-shows," before movie theaters became stationary (Christie 51). But capital for production was to be found in cities, and more profit could be made by locating movie houses there because the urban population had ever more expendable income and leisure time.

Cinema influenced the façades and topography of cities. So-called arcade "parlours" were one venue for regular film screenings; they carried "peep-show machines," which were viewed individually, and which offered a different viewing experience from the collective one of projected films (Christie 51). It was the projected pictures that necessitated buildings designed specifically for showing films, which started around 1905. Called "Nickelodeons," they had fewer than 200 seats to avoid theater taxes and were aimed at the lower classes and immigrants (Christie 51). Some years later cinema sought to appeal to the middle class by changing the content of films and constructing lavish theaters. In Paris, Moscow, and Berlin such theaters, which included orchestras and extravagant interior and exterior designs, became the new palaces of modern entertainment for the urban leisure class. As we will see in Chapter 1 using the example of Berlin, these movie palaces became the subject of sociological and philosophical debates during the 1910s and 1920s.

Even though Paris was not the only city associated with the early development of film, it was practically and symbolically an important site. Urban reconstruction turned Paris into an emblem of modernity when it was reconceptualized and redeveloped under the auspices of Baron Georges-Eugène Haussmann, who famously transformed the city from an organically grown town to a planned metropolis in the mid-nineteenth century. This new Paris took account of modern technology, such as railroads and gas lamps, and enabled the traffic to flow on grand avenues that were linked to the train stations. The buildings and straight, planned streets were characterized by a uniformity that had not been seen before. The kind of cityscape that Haussmann envisioned and executed characterizes Paris even today, including that signifier of modernity, the Eiffel Tower, which represents the world city par excellence. Even films of the French New Wave, in the late 1950s and early 1960s, use the Haussmannian cityscape to capture an authentic experience of Parisian urbanity.

Haussmann created a vertically organized city, in which the underground world of sewer systems and later subways embodied a hidden modernity which found its way into films about cities. This vertical organization took on symbolic and metaphoric significance for films beyond those set in Paris, as we will see in Fritz Lang's *Metropolis* (1927), Carol Reed's *The Third Man* (1949), and Richard Donner's *Superman* (1978), which map ideological values and/or class structure onto the urban structure of upper and lower worlds.

Urban sites – such as the street, the skyline, the bar – were important markers of cities in early cinema. The city street was a particularly privileged setting for action in early cinema. Many city films integrated shots of city streets as a recurring motif without advancing the narrative. Again, such scenes connect diverse films from different periods and national cinemas, including Walter Ruttmann's *Berlin: Symphony of a Great City* (1927), François Truffaut's *The 400 Blows* (1959), Pier Paolo Pasolini's *Accatone* (1961), John Schlesinger's *Midnight Cowboy* (1969), Perry Henzell's *The Harder They Come* (1972), Ali Özgentürk's *The Horse* (1982),

Wong Kar-wei's *Happy Together* (1997), and R'anan Alexandrowicz's *James' Journey to Jerusalem* (2003). All of these films from Germany, France, the United States, Jamaica, Turkey, Taiwan, and Israel are characterized by repeated shots of city streets – in Berlin, Paris, New York, Kingston, Istanbul, Tel Aviv – and in each one the street becomes an important site to circumscribe urban space and to negotiate characters' subjectivity.

In Chapter 1 we will see that the street is often coded as a site of danger and sexual encounter, which in Weimar cinema was routinely embodied by the figure of the streetwalker, the female prostitute. The streets and the screens of the metropolis promised erotic possibilities that linked the city and cinema in the collective imagination. An emblematic example that prefigures the reworking and rewriting of these early motifs throughout the twentieth century is Edison's *What Happened on 23rd Street, New York City* (1901), later echoed when the wind above a subway grating blows Marilyn Monroe's skirt up to her waist in Billy Wilder's *The Seven-Year Itch* (1955) (Christie 49). The erotics of the street is a recurring theme from the early Weimar street film (Chapter 1) to the sexualization of the metropolis in contemporary gay and lesbian cinema (Chapter 8).

Modernity

Though as we have seen the so-called train effect disavows the complex roots and inconsistent developmental trajectory of early cinema, scholars have returned to this founding myth as key to its relationship to modernity, which was experienced as a shock in the West. Tom Gunning, for example, interprets the myth, but does not take it as an accurate description of what happened:

> The on-rushing train did not simply produce the negative experience of fear but the particularly modern entertainment form of the thrill, embodied elsewhere in the recently appearing attractions of the amusement parks (such as the roller coaster), which combined sensations of acceleration and falling with a security guaranteed by modern industrial technology.
>
> (1989: 37)

Gunning suggests examining the train effect for "its metaphorical significance and irrational appeal" (2006:19), because the moving train embodied the changing perception of time and space in modernity – space as urban versus rural and time as modern versus premodern. Films manipulate space and time, whereas trains collapse space and require the concept of universal time. Until the advent of railroads, time had been local, often differing from village to village, but with the invention of the train it had to become consistent across space. Christie suggests that "trains, timekeeping and moving pictures all came together at the turn of the century to create a new image of time" (32). Time and space were becoming increasingly abstract, a feature they shared with other aspects of modernity (see Clarke and Doel), and film provided a venue for working through these concepts and their far-reaching consequences. So it is not surprising that moving trains are important in films that are emblematic of modernity, such as Walter Ruttmann's *Berlin: Symphony of a Great City* (1927), that mark important historical moments, as in Wolfgang Staudte's *The Murderers Are among Us* (1946), and that take on an allegorical historical function, as in Lars van Trier's *Zentropa* (1991).

Important for film as a new medium associated with modernity was also the filmic construction of those unable to negotiate the city with its pitfalls and its pleasures. Thus, early films often featured the figure of the country bumpkin – "the rube" of American vaudeville – who enters the city and is unable to read its clues appropriately, finally becoming the object of a crime or reacting foolishly to a film. Such stories posited an imaginary film audience that, unlike such characters, was urbane enough to negotiate cities and cinema successfully. The trope of the approaching train on celluloid became a playfully rhetorical figure, which separated the urbane, film-going public from the terrorized country bumpkin incapable of comprehending the new medium. Christie describes a "1901 film by Robert Paul, in which a bumpkin tries to look 'behind' the screen on which he has seen an approaching train, and succeeds in pulling the sheet down," contrasting this with a "British story from 1904 entitled 'The Cinematograph Train'," in which young Bobbie sees a train rushing towards him in a cinematograph show and steps onto the platform and into the train and rides off (13).

These narrative constructions of a rural character coming to the city unable to negotiate its dangers and seductions, often embodied by a female character, continue throughout the history of film. Of course there is truth to the experience of moving from the country to the city and being overwhelmed by the onslaught of stimulation, but it is important to realize that the idea of the sophisticated urbanite, able to engage appropriately with the pleasure of film, is also a filmic construction itself, one that grew out of economic interest in assimilating masses into consumers of a product.

The nation and national cinema

Paradoxically, the early history of cinema was strongly anchored in national contexts, even while it was characterized by international exchange. Only now, with globalization, are films commonly funded by more than one nation and distributed around the globe. Nations played important roles in the development of very early cinema even though one could not yet call it "national cinema."

Because of their technological innovation, the French studios Gaumont and Pathé were early leaders, and because silent film could be understood across linguistic barriers, Gaumont could open branches in London, Berlin, Moscow and New York (see Christie 93). Then Gaumont came under the control of MGM in 1924, indicating the end of French dominance in cinema and the beginning of American economic hegemony over the film industry, which continued throughout the twentieth century. Meanwhile, after the First World War Germany became the new force in Europe (see Chapter 1).

Although the story of cinema is embedded in different national contexts, the terms "national cinema" and "nation" are understood and defined variously by scholars. For some, nations are entities that exist prior to cultural expression and then are articulated through culture, while others propose that "nations are constructed in a process of myth-making linked to the needs of the modern, industrial state" (Hjort and MacKenzie 1). "Nation" is a key term for this book, largely because the nation provides the economic and political framework within which films are produced.

Much discussion of cultural production, particularly with regard to literature and film, relies on Benedict Anderson's study of nation as an "imagined community." Anderson argues that with the advancement of print, the novel offered a narrative form that allowed members of a particular nation to imagine themselves as belonging to the same nation despite geographical distance and lack of connection to other individuals of the same nation. The concept of the "imagined community" suggests that film, too, has played a particularly pivotal role in the ongoing development of national identities. It can also be applied to cities and neighborhoods, as well as to communities of revolutionaries, ethnicities, and gays and lesbians as in films discussed here.

Cinema has developed from national cinemas to transnational cinematic practices as a result of globalization, which has reduced the power of the nation state. Increasingly filmmakers are trained abroad, receive multi-national funding, and make films for a world market, and increasingly narratives involve characters that travel across borders. In the early development of cinema national capitals, such as London, Paris, Berlin, and Moscow were at the forefront of the development of the new technology. Only the American film industry created a place at a distance from the nation's capital that became established as the capital of filmmaking – Hollywood. So the first three chapters here trace the history of the city film from Berlin via Hollywood to Paris and lay the foundation for the rest of the book. The remaining chapters address different kinds of cities associated with specific thematic concerns, but show the move from national cinemas to transnational cinematic practices.

Hollywood and the studio system

The studio industry and independent filmmaking are two poles in the organization of filmmaking. American cinema has come to stand for the studio system – though it includes both studio and independent cinema – and European cinema has come to stand for independent cinema – though most European national cinemas also rely on studio production. Film production can fall into either category or integrate both in a mixed form. The development and solidification of the studio system coincided with the feature narrative form as we know it today, around 1912–13.

The vertically integrated studio system, which refers to simultaneous control over production, distribution, and exhibition, began in France in 1910 with the three production companies Gaumont, Pathé, and Éclair (Hayward 354). Nevertheless, it is primarily associated with Hollywood and traditionally dated around 1920. In the Hollywood studio system a production head supervises the production, which is characterized by a division of labor and mass production of films, which are shot out of sequence. In 1917 Adolph Zukor vertically integrated the studio system when he bought the distribution company Paramount Film Corporation and connected it to his own production company, Famous Players–Lasky Corporation, which led to his control over production and distribution (Hayward 354). In the 1920s the five major studios – Paramount, Fox Film Corporation, Metro– Goldwyn–Mayer, Warner Brothers, and RKO – became fully vertically integrated, while Universal Pictures, United Artists, and Columbia, in contrast, did not own theaters but used the majors' theaters.

From the 1920s to the 1930s studios monopolized the film industry, increasingly organizing production in different departments relying on specialists. From 1930 to 1948, Hollywood's studio system dominated the field with different studios, each of which created its own look by having its own stars, scriptwriters, directors, and designers. Closely associated with the studio system are genre films, in which the content is organized according to recognizable types which are defined by conventions, like the Western, the musical, and the melodrama. These genres, however, are not static; because they reflect audience expectations, they can change over time and they can be combined.

Globalization and transnational cinema

By making borders increasingly permeable to capital and commodities, globalization is a force that has substantially increased the global exchange of goods, including cultural products. At the same time, however, borders have been increasingly closed to people, who often face the violence and exploitation associated with illegal migration. Electronic communication and digital culture are also increasingly detached from the nation-state and neither invoke a country of origin nor address a national audience in an explicitly recognizable framework of national culture. Taking stock of transnational cinema, Elizabeth Ezra and Terry Rowden describe the difficulty of assigning "a fixed national identity to much cinema," noting that the "stable connection between a film's place of production and/or setting and the nationality of its makers and performers" does not exist anymore (1). They outline the substantive adjustments that cinema and new media have undergone with the changes in transnational production, education, and reception of films and filmmakers.

Transnational cinema includes Hollywood's domination of global markets, other transnationally distributed films such as those from the Hong Kong film industry, collaborations between former colonial countries and European countries, which fund many African, Caribbean, and African-American films, and European co-productions. Both dominant and subversive cinematic cultures circulate in transnational structures of funding, training, and distribution. This means that independent filmmakers also have access to web and digital distribution networks, even if they are economically disadvantaged (*vis-à-vis* Hollywood). Chapter 9 approaches the cinematic representation of global cities on the basis of a range of films that show movement across different kinds of national borders. As will become clear, the cosmopolitan and metropolitan city is of transnational importance in the development of globalization in a paradoxical way: on the one hand, the megalopolis as detached from the nation state is increasingly important while, on the other, the importance of cities as sites for labor decreases with globalization.

Postmodernity

There are competing definitions of modernity and postmodernity. The coherence of modernity relies on its tie to modernist art, architecture, urban planning, and design; most scholars define postmodernity as a reaction and contrast to modernity. Theorists of postmodernity see it as a prevalent cultural and political phenomenon related to late capitalism and philosophical traditions that are critical of the European Enlightenment heritage represented by, for example, Immanuel Kant who believed reason to be the ultimate authority, and Karl Marx, who propounded the betterment of all through class revolution. Jean François Lyotard argues that such metanarratives have become meaningless in view of the power of capital (120).

Fredric Jameson sees the "postmodern" most clearly in relation to cities on the one hand and war on the other, suggesting that the term is also supremely applicable to cinema which, like postmodernism, is very much tied up with representation. As with the advent of modernity in the early twentieth century, time and space have also undergone change in postmodernity. Jameson explains postmodern architecture as "a mutation of built space itself," suggesting that while modernist architecture was utopian, the new architecture of postmodernity is not (10). His famous example of the Westin Bonaventure Hotel in downtown Los Angeles, which was built by architect and developer John Portman (11–16), creates a total world indoors, like contemporary malls, hotels, and train stations that include shops and restaurants. Much of what Jameson describes as characteristic of individual postmodernist buildings also applies to the postmodern cinematic portrayal of architecture and cities, especially in films with dystopian visions of the future such as Ridley Scott's *Blade Runner* (1982), Andy and Larry Wachowski's *The Matrix* (1999), and Alex Proyas's *Dark City* (1998) (see Chapters 5 and 6). Jameson adds that postmodern narratives about war exceed the "traditional paradigms of the war novel or movie" and reflect a "breakdown of any shared language" (16). He contrasts the movement of the locomotive, representative of the modern machine, to the postmodern machine, which "can only be represented in motion" (17). The staging of the rebel ship *Nebuchadnezzar* in *The Matrix*, which we never see moving through space, echoes his characterization of postmodernism.

This shift from modernity to postmodernity as a shift from metanarratives to fragmented explanations becomes clear in a comparison of modern and postmodern science fiction. Fritz Lang's *Metropolis* (1927), for example, uses the conflict between Christianity and socialist politics to posit a solution to the exploitation of the working class, relying on the assumption that the audience can recognize and integrate both sides in their belief system. In contrast, postmodern films such as *Blade Runner*, *Dark City*, and *The Matrix* do not rely on such metanarratives, but rather offer fragmented, individualized theories advanced by individual characters that do not cohere into a sustained narrative system applicable to a universal argument.

Such narratives reflect another aspect of postmodernity, one identified by Christopher Butler: "suspicion which can border on paranoia" (3). All the films discussed here as examples of postmodern cinema share a paranoid narrative and paranoid characters, which in turn lead to another important aspect of postmodern theory. Jean Baudrillard proposes that in contemporary media-saturated society, representation substitutes for real, that is, in contemporary society media simulate reality so convincingly that the audience becomes more familiar with the simulacrum than with the real. Thus, while in early cinema the difference between the real and the cinematic train existed, Baudrillard argues that today "the real and the imaginary are confused in the same operational totality" (187). A striking example is that for people who have seen Steven Spielberg's *Schindler's List* (1993) the reality of Auschwitz may be subsumed in the filmic representation – that is, visitors to the actual concentration camp may well spontaneously think that the gate to Auschwitz looks just like the one in the film rather than the other way around. This does not mean that Auschwitz did not happen or that visitors cannot distinguish between the real and the mediated site, but it illustrates the anxiety expressed in postmodern science fiction films that the real and the mediated may become indistinguishable.

Ironically, while these postmodern films stage an inability to discern between reality and representation, digital imaging processes can create virtual reality, portraying things that are not and have never been real. Country bumpkins in early films often could not differentiate between the world of the screen and reality, but the basic assumption was that a clear difference existed between the two, and that assimilation into the sophistication of an urbanite would allow spectators to make those distinctions. In contemporary postmodern film, that distinction has disappeared and the narrative is characterized by the desire and search for it. The heroes of postmodern science fiction films are the urbanites of yore, who can distinguish between the real and the simulacrum and can control the virtual environment. And like those early sophisticates who were used to advertise films, these contemporary figures become ideal literates of a new medial revolution used to sell computer games, in which spectators can act out control over the virtual environment. The examples of global cinema in this book's final chapter are not characteristically postmodern, though in specific cases they portray a postmodern world. Though global and transnational, the transnational films discussed in Chapter 9 emphasize local and individual experience and argue implicitly against the simulacrum advanced by Hollywood cinema.

History is accorded a different role in cultural production in postmodernity than in modernity. Rather than a reference point for locating the action in a precise moment in time, however, history is turned into an archive from which films cite, often mixing and matching incongruous references. This postmodern tendency is also evident in contemporary art and design, which in turn constitute the setting for postmodern films. Jameson emphasizes the importance of the "pastiche," a term that refers to a work of art or literature that integrates references without creating new meaning. A pastiche can thus be temporally confusing, implying even that history has become irrelevant. For example, in postmodern films on war (discussed in Chapter 5) the narratives reference the Second World War but cannot be connected accurately to a specific time and place, for example in Marc Caro and Jean-Pierre Jeunet's *Delicatessen* and Lars von Trier's *Zentropa* (both 1991). Postmodernism therefore emerges from its own historical moment, even though it creates the illusion that it is beyond history by irreverently quoting from different historical periods, as well as cultures and styles.

Urban and cinematic space and temporality

Analysis of both film and the city involves the coordinates of space and time, which according to Edward W. Soja have traditionally been seen as juxtaposed, with either history or geography in the ascendant (9). But David B. Clarke's focus on the relations between "urban and cinematic space" notes that conceptions and constructions of space connect cinema and the city (2). Because spatial categories of analysis – topographies, sites, settings, locations – change throughout history, Soja integrates a critical understanding of history and geography without privileging either. He analyzes "cityspace" following Anthony Giddens's concept of "spatio-temporal 'structuration'," which links social structures – the family, the community, the social class, the market economy, the state – to the "dynamic production and reproduction of cityspace" (9). When films cinematically construct space to mark social class and cultural developments, they rely on the knowledge and recognition of the audience. Like cities, films engage in processes of production and reproduction of social relations in spatial configurations. Henri Lefebvre conceptualizes "the relations between spatiality, society, and history in a fundamentally urban problematic," and argues that the social relations of class, family, community, market, or state power "are specifically *spatialized*," meaning that social relations are translated into "material and symbolic spatial relations" (9).

How to read a city?

The city has always been particularly important in understanding how social change manifests itself. And urban studies has begun to address films as cultural visions of what cities represent because as Mark Shiel and Tony Fitzmaurice point out cinema is "a peculiarly spatial form of culture" (2001: 5). John Rennie Short's *Urban Theory* implies that "modernity, capitalism and postmodernity" link the study of film and the study of cities, and in his overview of ways to read a city he particularly emphasizes the "operation of power and the struggle for power," which organize the city and are reflected in it (2–3). He suggests that these power relations are organized by social differences in class, gender, age, race, and ethnicity, which produce urban patterns and processes. Films reflect such urban patterns in how they code neighborhoods as rich or poor or landscapes as urban or rural. They reflect class in costume and setting, and in whether characters are positioned inside elaborate domestic spaces or outside in the urban public space.

How to read a film?[1]

An analysis of a filmic representation of a city begins most helpfully with observing how individual films represent the conditions of said city or neighborhoods in the specific historical moment, and then moves beyond seeing film as mere representation of social reality to focus on how the cinematic text constructs and comments on those conditions. Film scholars are increasingly paying attention to the construction of space. In contrast to an actual city or urban space, however, any kind of filmic space is mediated, and there are several levels of analysis to take into account:

1. what film shares with drama: *mise-en-scène*, which consists of setting, acting, costume, and lighting;
2. cinematography – the actual manipulation of the film strip in the camera and in post-production (Bordwell 477);
3. editing, which creates continuity or discontinuity with regard to both space and time, and creates the speed and pace of a film, a scene, or a sequence – often a significant cultural consideration;
4. shots, the uninterrupted, continuous movements of film, which are connected to each other during the editing, such as an opening long shot of a well-known city skyline to establish the general setting, or sequences which juxtapose urban and rural, life and death, rich and poor, old and young, and so on;
5. sound, which came of age in 1927 but had been experimented with much earlier, and which can be divided into noise (sound effects); music, which can be diegetic (part of the story) or non-diegetic (sound track), on or off screen; and speech, which can be subjective (in a character's mind) or objective (so that the audience can hear it).

Cities and Cinema discusses the different ways in which cities and cinema intersect, and in so doing relies on resources and methods from both urban studies and film studies. Ultimately, however, the latter is emphasized to highlight the importance of cinema in understanding how cities are imagined. The objective is to examine how these imaginary cinematic cities work through issues of the national and the global, modernity and postmodernity, and the reproduction of power associated with race, class, gender, and migration. This book discusses films that not only reflect and comment on urban issues but also create cinematic visions of cities beyond of what we experience in our daily lives.

Further reading

Early film history and modernity

Ian Christie (1994) *The Last Machine: Early Cinema and the Birth of the Modern World*, London: BBC–BFI. A well-written and well-organized introduction to early cinema that makes a convincing argument about the relationship between cinema, the city, and modernity.

Thomas Elsaesser with Adam Barker (1990) *Early Cinema: Space – Frame – Narrative*, London: BFI. A collection of essays by authorities on early cinema organized according to the changes in space and time, the economics of the industry, and the development of editing.

National cinema

Mette Hjort and Scott Mackenzie (eds) (2000) *Cinema & Nation*, London: Routledge. An edited collection that engages with different approaches to the concept of *national cinema* with examples from different national contexts and relying on different sociological, historical, and aesthetic approaches.

Transnational cinematic practices

Elizabeth Ezra and Terry Rowden (eds) (2006) *Transnational Cinema: The Film Reader*, London: Routledge. This collection brings together previously published essays that have advanced the discussion on transnational cinema for an overview of the diverse and important approaches to the development from national to transnational cinema.

Postmodernism

David Harvey (1989) *The Condition of Postmodernity: An Enquiry into the Origins of Cultural Change*, Malden, MA: Blackwell. Harvey explains postmodernism from the perspective of architecture and urban design.

Essential viewing

The Movies Begin (1894–1913, DVD Box Set, Video King). Includes shorts by Edwin S. Porter, Thomas Edison, Louis Lumière, George Méliès, Alice Guy Blanché, D.W. Griffith, and R.W. Paul.[2]

SECTION I

This section establishes the significance of the relationship between cities and cinema with three chapters that chronologically pair cities with groups of films. Chapter 1 describes the emergence of the city film during Germany's Weimar Republic, its first and ill-fated democracy, from 1919 to 1933, particularly in relationship to Berlin, the country's capital and cultural metropolis. Chapter 2 relates the cycle of film noir, highly stylized, black-and-white, American, private-eye films, to the immediate post-Second World War period, establishing Hollywood as the location for film production and Los Angeles as the films' primary setting. Chapter 3 focuses on the relationship of the French New Wave – a group of filmmakers who rebelled against the French studio system during the late 1950s and early 1960s – to Paris, the topic and the site of production and reception of their films. Taken together, these three chapters provide a historical outline of the interconnected development of cities and cinema from the 1920s to the 1970s.

Cities have been central to the development of cinema in its three central aspects: production, representation, and reception. Individual cities are important sites for film production, functioning thematically and providing settings for stories as well as sites for their distribution and consumption. Because Berlin, Hollywood, and Paris signify different modes of film production, together they tell its story: Weimar Berlin stands for the different early organizational forms of film production; post-Second World War Hollywood represents the studio system at its height; and Paris of the 1960s exemplifies *auteurism* – an emphasis on the vision of the individual film director.

Section I, then, lays the foundation for this book's overarching point: that the connection between cities and cinema lends itself particularly well to demonstrating larger cultural shifts from national to transnational and modern to postmodern contexts. Chapter 1 illustrates how modernity was simultaneously experienced as violent shock and embraced for its technological and aesthetically innovative opportunities. This Janus-faced quality continues in the portrait of the modern city in film noir and the French New Wave. The cinematic vision of the modern metropolis emphasizes its public spaces, such as streets, in contrast to rural and suburban spaces.

The three chapters address three classic examples of city films but also portray famous moments in national cinemas: film noir embodies American film, while Truffaut and Godard represent Frenchness *par excellence*. Each city is representative of its nation and each group of films represents a defining moment for its national cinema. At the same time, the portrayal of the cities points to concepts – modernity, urban alienation, human desire – beyond their respective national dimension.

The three groups of films are also linked by the influence they had on each other: style and themes of Weimar cinema reappear in film noir, which in turn is quoted and referenced by the directors of the French New Wave. These connections do not contradict the understanding of national cinema; even during its height, transnational exchange was common, sometimes resulting from migration, as when filmworkers from the Weimar Republic fled the Hitler regime and found work in Hollywood. Later, French directors paid homage to individual filmmakers and actors in Hollywood who sometimes, like Fritz Lang, were film émigrés from Germany.

Section I provides a roadmap to these cinematic exchanges that demonstrate continuity in the filmic representation of cities in the form of topical and aesthetic similarities. Weimar cinema is most famously associated with Expressionism, a highly dramatized, black-and-white cinema emphasizing shadows and the play of light and dark, in which realism was subordinated to the vision of an individual director. These aesthetics can be traced from Weimar cinema to film noir and to the French New Wave. Similarly, but less recognized, Weimar's urban cinema also often relied on New Objectivity, an art movement that represented modern life realistically, which also reappears in the films about Los Angeles and Paris.

Finally, gender cuts across these three cinematic moments, projecting dangerous and sexualized femininity in the public space of the city. The streetwalker of Weimar cinema is reincarnated in the femme fatale in film noir and lives on in the woman as unreliable object of desire in the French New Wave. These figurations of femininity are accompanied by male characters who either master the urban space or are seduced by the sexualized and dangerous urban woman, often interchangeable and conflated with the city in question.

1 Modernity and the city film: Berlin

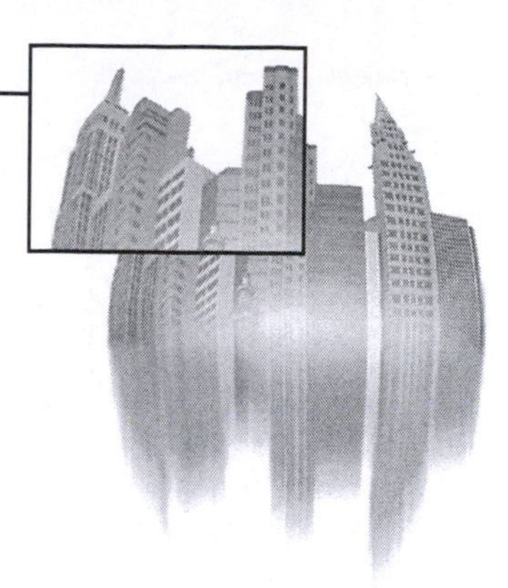

Today all segments of the population stream to the movies, from the workers in suburban movie theatres to the haute bourgeoisie in the cinema palaces.

Siegfried Kracauer

Learning objectives

- **To understand the role of Berlin in the development of the city film in the Weimar Republic**
- **To outline the concept of modernity and the contributions of its different theorists**
- **To define the genre of the street film and consider its gendered dimensions**

Introduction

"A new genre was born: 'city film'," claims Helmut Weihsmann about *avant-garde* films in the mid-1920s (10). City films as a crucible of modernity created urbanity as the modern space, and during the 1920s in Europe, this modern city *par excellence* was Berlin. After the First World War Berlin played a central role in Germany and Europe as the locus of modernity and cosmopolitanism, a place where modernist art flourished. The city of Berlin was the theme of several city films, the site of production with several studios located on its outskirts – including the famous Ufa in nearby Neubabelsberg – and it was also the site of elaborate movie theaters where important premières took place. Berlin was also a place of coffee-houses, bars, newspapers, and magazines where those who entered the new industry met and networked, and those who wrote about the city and its culture gathered to discuss. Berlin was central to the development of the cinema and from its inception German cinema has been "preoccupied by the big city as a site of adventure and modernity" (Kaes 1996: 65).

The birth of the city film

Weihsmann suggests that the early filmic depiction of cities in the 1920s resulted from a growing fascination with "metropolitan motifs, motion, and development" and from the assumption that the camera could capture visual evidence of a city. In "documentary style" city films filmmakers reproduced different "urban motifs," while in "pictorial *colportage*" they mixed documentary footage and fiction shot on location (9). Karl Grune's *The Street* (1923), Friedrich Wilhelm Murnau's *The Last Laugh* (1924), G. W. Pabst's *Joyless Street* (1925), Fritz Lang's *Metropolis* (1927), Walter Ruttmann's *Berlin: Symphony of a Great City* (1927), Robert and Curt Siodmak's *People on Sunday* (1928), Joe May's *Asphalt* (1929), and Lang's *M* (1931) were all produced within one decade, and all take place in a city – most often Berlin, but not always – and thematize urbanity, especially the

Figure 1.1 ***The Last Laugh*: The city at night**

period's understanding of the dangers and pleasures of modern urban life: crime, anonymity, a loosening of morality, unemployment, and class struggle on the one hand, and movement, speed, entertainment, and liberated erotics on the other. These films foreground what David Frisby has identified as characteristics of modernity: "abstraction, circulation and movement and monumentality" (20). While some were fascinated by the cinematic possibilities of documentary realism, others were fascinated by the artificiality of the set. The latter is famously the case with Fritz Lang's *Metropolis*, but also with *The Street*, for example, which has a highly artificial set.

The genre of the city film thus integrates the aesthetic and the documentary aspects of film. It constitutes a genre with its own history and a prism through which to address a host of related and interconnected topics regarding cinema and urbanism. Several films depict the city as the setting for social problems: *M* famously tells the story of the search for a child murderer, and *The Last Laugh* portrays the fate of a hotel employee who has lost his position but continues to wear his uniform to garner respect. The figure of the prostitute embodies both liberated and commodified sexuality located in the streets of the metropolis, for example in Grune's *The Street*, Pabst's *Joyless Street*, and May's *Asphalt*. All these films except for *The Last Laugh* constitute the genre of the street film developed between 1923 and 1925.

Other social problems, such as class conflict, perceived as crucially defining the urban metropolis in the early twentieth century, are expressed spatially, as in Lang's *Metropolis*, where a vertical, futuristic city is segmented into the upper world of the factory-owner and the lower world of workers, portraying a dystopian vision of urban modernity. Images, events, and encounters seen as representative of urbanism become the raw material for *Berlin: Symphony of a Great City* and *People on Sunday*. Most importantly, however, the experience of the modern metropolis changed visual perception and yielded new narrative forms and possibilities for aesthetic representation: abstract shapes and compositions, episodic narratives, and cinematic montage express the experience of urban modernity. Because film was part of the newly emerging mass culture, changing in tandem with urban modernity, cinema of the 1920s functioned in a double role as both "product of urban modernity" and "producer of urban culture" (Weihsmann 10).

The film industry created not only artificial cities as settings for films, but also an artificial city for film production: Neubabelsberg, in the no-man's-land between Berlin and Potsdam. The Weimar Republic witnessed the early development of the studio system, particularly with the growth of Ufa (Universal Film Aktiengesellschaft), the studio that Klaus Kreimeier labels "one of the most important movie studios in the world" (3). It was founded during the First World War for the purpose of creating national propaganda. In this "film-city" (Ward

Figure 1.2 ***The Last Laugh*: A typical Berlin working-class tenement court yard**

2001a: 21), unemployed and underemployed architects created their architectural visions in set designs (22), because in the immediate post-war period building projects were denied to them (28). Janet Ward emphasizes the artificiality of Neubabelsberg, describing it as "Babelesque, consisting of towers and tunnels over eighty-odd acres of artificially lit outdoor and indoor playgrounds" (21). And while movie production created its own fantastic city outside of Berlin, moviehouses, called film palaces, changed the face of the metropolis itself.

In Germany the rise of modernity was accompanied by theoretical discussions articulated by Siegfried Kracauer, Walter Benjamin, Georg Simmel, and Max Weber, who had a keen interest in the city and the cinema. Their theories offer us ways to think about cinematic representations of urbanity with regard to the city film from the Weimar Republic, but they also provide us with foils for discussions about the cinematic representation of urban space in general. The next section provides an introduction to these crucial figures in their respective contexts. Their contributions to the theorization of urbanity remain reference points throughout this book.

Theories of modernity and urbanity

Sociologist Georg Simmel noted the importance of the emerging metropolis for changing life, culture, and subjectivity in the early twentieth century. His seminal essay "The Metropolis and Mental Life" (1903) focuses specifically on the effect of the city on subjectivity and describes "the metropolitan type" as characterized by a rational and intellectual response to uprootedness, the increased speed of information and impressions, and the "*intensification of nervous stimulation*" (175, italics in the original). Simmel noted that in contrast to the quiet life in rural communities characterized by social networks, kinship, and family, the metropolis produced quickly changing impressions on individuals. Discontinuity and fragmentation characterized city life, where actions and events assaulted individual inhabitants actively and unexpectedly. The shift at the end of the nineteenth and the beginning of the twentieth century created a radical rupture of the "sensory foundations of psychic life" and created a new kind of "sensory mental imagery" (175). Simmel described the effect of the modern metropolis on subjectivity in a combination of imagery and sensory perception, motion and stimuli, a combination that encapsulates the potential of the medium of film to express the characteristics of the city. A film such as *Berlin: Symphony of a Great City* reproduces the sensory experience of the city through its "associative montage," a method which can capture the fragmented aspects of modern life in the metropolis (Hake 1994: 130).

Other films combine the contrast between the rural and the urban environment with a developmental narrative from rural to urban. German director F. W. Murnau's *Sunrise – A Song of Two Humans* (1927), made in the United States, is a case in point. A woman from the city tries to seduce a man from the country to kill his wife while crossing the river on the way to the city. The city is embodied by the destructive seductress and the country by the wife, who is also a mother, a caring and quiet character defined by her social roles in the rural village. In contrast, the woman from the city is characterized by her independence and her appearance: clad in a sexy black dress, smoking, and using make-up, she is the incarnation of "the archetypal metropolitan female of the 20s" (Fischer 43). If the husband does indeed drown his wife, the familial ties connecting him to the rural soil will be destroyed. Instead, however, he travels to the city with his wife and experiences with her its pleasures and dangers. On their arrival, echoing Simmel's description, traffic surrounds and overwhelms them and the wife is almost run over. They find many distractions – going dancing, having their photo taken, and visiting a barber shop.

Sunrise also refers to the role of cinema in negotiating the contrast between country and city, generally associating cinema with the latter. When the city woman tells the man about the city, a film is projected against the rural sky transposing the city onto the rural environment. Once the man and his wife are in the city, they have their photo taken in a studio, thus participating in the modern technology of visual self-representation. They are adding their technological and visual literacy to the journey, which becomes a passage into maturity associated with the city. Raymond Williams proposes that the idea of the country is paradigmatically associated with "childhood" (297), and thus the individual story of the couple from the country represents the history of humankind, outlined by Simmel as a shift from rural to urban. The city woman attempts not only to destroy the relationship between the husband and his wife, but also to sever his ties to the land by persuading him to sell his farm and take his cash to the city. This narrative movement ties money economy to the city and portrays the emotions expressed by the city woman as calculating rationality, a construction that also reflects Simmel's theories regarding the metropolis and Max Weber's discussion of rational capitalism, which I return to later in this section.

The metropolis is defined by Simmel as a place of money economy, which for him goes hand-in-glove with the metropolitan rationality that redefines human relationships in terms of exchange value and turns all action in the metropolis into "production for the market" (176). The potentially alienating effect of the metropolis necessitates that the metropolitan character reacts with "his head instead of his heart," which we will see demonstrated in a case study of *Metropolis* (176). In order to be most efficient and highly productive, according to Simmel, cities display the "highest economic division of labor" (182).

This emphasis on the division of labor is reflected in many films that depict differentiated professions, from *Berlin: Symphony of a Great City*, *Sunrise*, *M*, and *The Last Laugh* to its most explicit articulation in Lang's *Metropolis*, which famously shows a city in which the class relationships are expressed through its spatial compositions (see the case study at the end of this chapter). The owners of the means of production live and socialize in spacious offices, gardens, and a sports arena, all of which are elevated, while the workers of Metropolis live below the surface with no access to light, art, or nature, and are reduced to their functions in a differentiated workplace. Frederson, the boss of Metropolis, is characterized as the head, the workers as the hands, and the film moves towards establishing a triangulation in which a representative of the heart connects the fragmented aspects of production that would otherwise be alienated from each other.

Simmel describes characteristics that we find not only in the early city film but in the visual and narrative strategies of present-day urban films, such as the emphasis on one well-known city that is inhabited by an urban type. He differentiates between cities that have become significant through "individual personalities" – such as Weimar, which will be forever associated with Johann Wolfgang von Goethe – and the metropolis, which is significant in and of itself, even beyond its physical boundaries (182). This explains the predominance of Berlin as the setting in the Weimar Republic city films. And the particular urban types that we find in city films from different time periods and geographical locations echo Simmel's descriptions: there is first and foremost the "blasé attitude" that he ascribes to characters shaped by the urban experience. In the films analyzed in this book, we often see the figure of the migrant arriving in the city and encountering this blasé attitude. Simmel's theoretical account, "The Metropolis and Mental Life," evinces the ambivalence of artists and theorists about modernity. While his description of the metropolis is in many ways critical, he also acknowledges the freedom of the individual in the city, which can, however, have the effect of feeling lonely and lost in the metropolitan crowd.

The two crucial moments articulated by Simmel – movement through the city, and the commodification of relationships in the city – were extensively theorized by Walter Benjamin, the foremost philosopher and cultural critic during the Weimar Republic. He observed and described the *flâneur*, strolling leisurely through the city, as a key figure in nineteenth-century Paris and then in Berlin in the early twentieth century (see the lyric poetry of Charles Baudelaire for a literary account of the *flâneur*). As Jaimey Fisher points out, Benjamin returned to the figure of the *flâneur* in the 1920s when he reviewed Franz Hessel's *On Foot in Berlin* (*Spazieren in Berlin*) (461). The *flâneur* was the idle person of the nineteenth century who wandered the city aimlessly and sought "refuge in the crowd" (Benjamin 1999a: 21). Benjamin points out that around 1840 it was considered elegant to take turtles on walks through Paris, and the turtles determined the pace of the *flâneur* on his stroll through the city.

In contrast to Simmel's model, however, the crowd in Benjamin's understanding was a veil "through which the familiar city beckons to the flâneur as phantasmagoria," creating quite a different relationship between the crowd and the individual from Simmel's and giving us another understanding of the use of crowds in city films: as a reference to the cinematic fantasy and variety of the city (1999b: 10). Benjamin recognized the cinematic quality of the metropolis, and his own writing mimics the process and effect of editing and juxtaposing interior and exterior spaces when he describes urban space as "now a landscape, now a room" (1999b: 10). David B. Clarke explains that "the practice of *flânerie* and the apparatus of the cinema both changed the social meaning of presence, and did so in much the same way; both effectively embraced the virtual" (5).

In "The Return of the Flâneur" (1929) Benjamin mused about the rebirth of the *flâneur* in Berlin in the 1920s (1999c: 262–7) and concluded that the *flâneur* can read the past because he can recognize it from his perspective of modernity; the *flâneur* is a figure who emerges at the particular moment of modernity. Yet, the *flâneur* does not entirely belong to modernity; instead he is positioned "on the threshold – of the metropolis as of the middle class. Neither has him in its power yet" (1999b: 10). Susan Buck-Morss also sees a "utopian moment of flânery," but describes it as fleeting, because the conditions for flânery had already passed by the time Benjamin was writing (344). Like Simmel, Benjamin projected his own ambivalent relationship to the city and modernity onto the figure of the *flâneur*. On the one hand, he described the phantasmagoria of the modern metropolis, while on the other he preferred a mode of movement that belonged to a period that had already passed (Fisher 474–5).

Benjamin's description of the *flâneur* is still deeply ingrained and shaped by gender distinctions to the extend that Anke Gleber points out that the "female flâneur has been an absent figure in the public sphere of modernity, in its media and texts, and in its literatures and cities" (1997: 69). Benjamin aligned Baudelaire's Parisian *flâneur* with the asocial beings whose "only sexual communion" is "with a prostitute" (1999a: 21); in the mind of the *flâneur*, woman appears only as sexual commodity. The gendered binary is mapped onto the urban landscape: the *flâneur* scouts the marketplace of the city, and the woman, as prostitute, constitutes the city's sexualized commodity. Like Simmel, Benjamin emphasized the gaze, but only the gaze of the *flâneur*, who would lose himself in the crowd but nevertheless remain "the alienated man" (1999b:10). In Baudelaire's poetry, according to Benjamin, Paris becomes the site where woman intermingles with death, which creates an association between woman, death, and the city that recurs often in films about the danger of the city.

The triangle of woman, death, and the city is most clearly embodied by the femme fatale, who represents death and the city. This figure can be traced from Weimar city film (in such figures as the false Maria in *Metropolis*) to her incarnation in film noir as a character who seduces men to kill. Most often these femmes fatales emerge from the anonymity and chance encounters of the city. Benjamin used the fetish of the prostitute, seller and sold in one, to resolve the contradictions between inside and outside, modernity and the past (1999b: 10). In the literature, film, and early-twentieth-century German theory the figure of the prostitute functioned to negotiate female sexuality and gender relations, particularly in the city of Berlin. Jill Smith argues that her figure "destabilize[d] the traditional sexual roles of male and female – that of male agency and female passivity" (3).

Many of the important theorists on the interconnection of the city, cinema, and modernity emerging out of the 1920s in Germany were in dialogue with each other, as were Benjamin, Simmel, and Siegfried Kracauer. Both Benjamin and Kracauer were concerned with changes in perception, the emerging masses in the metropolis, and the subsequent changing character of art. In his famous essay "Work of Art in the Age of its Technological Reproducibility," Benjamin mapped out the historical changes in art, from painting that carries an aura to the cinema, which relies on technical reproducibility and creates distraction as a form of reception (2003a: 269).

Kracauer was also critical of the entertainment industry's mimetic claim, which, he argued, displaced history and sold daydreams to the newly constituted audience of female white-collar workers. He particularly reproached the "surface" culture produced by modernity, which he saw manifested in the Ufa studio in Neubabelsberg: "350,000 square meters house a world made of papier-mâché. Everything guaranteed unnatural and everything exactly like nature" (281). This surface culture was seen by theorists as produced by capitalism's attempt to turn content into a product for consumption. It was particularly Kracauer's training as an architect that provided him with a "method of approaching a subject from its surface structures" (Hake 1993: 263).

The disavowal of history in Kracauer's mind was enabled by the cutting apart of reality in the editing process, and creating and destroying the sets and props for films, and then reconstituting "a world out of these pieces" (287). The films that circulated, he rightfully observed (and later mapped out with regard to German film in his most influential study, *From Caligari to Hitler*) "are the daydreams of society" (292). Such films addressed themselves particularly to the new audiences of what he called "little shopgirls," created by the changing gender roles and new professions that accompanied new technological and urban developments (291–304). These young women had white-collar jobs outside the home as telephone operators, secretaries, and shopgirls. Kracauer's critique of the fantasies advanced in these films and his critical analysis of the Ufa studios was also mirrored in his critical assessment of the movie-houses as "palaces of distraction," characterized by "*surface splendor*" and constituting "shrines to the cultivation of pleasure" (323, emphasis in original). Such places aimed to entertain the masses who constituted the city, particularly Berlin.

Kracauer suggested that modernity in the city created the masses, a social organization that in turn shaped cultural manifestations in the city. Their influence moved from the exterior to interior; in Kracauer's words, the life of the street gave "rise to configurations that invade even domestic space" (325). The opening of Fritz Lang's *M* illustrates this understanding of the relationship between the exterior and interior urban environments. The film opens with children playing in a Berlin tenement courtyard, singing a song and playing a game about a murderer who, the audience finds out shortly, is terrorizing the city. The camera moves from the children's play in the courtyard to a working-class woman carrying laundry to the house and up the stairs in the otherwise empty staircase, a hybrid space between the public and the private, and entering the apartment of Elsie Beckmann's mother, who is worried about her daughter's failure to arrive home from school yet. The children's play mirrors her anxiety, and the camera's movement from the exterior to the interior reflects the movement of terror and anxiety from exterior, social to interior, psychological spaces.

Berlin became the representational metropolis, both in films such as *M* and in Kracauer's study of the relationship between the masses, the city, and distraction, represented by the cinema. He explained that Berliners were "*addicted to distraction*" as a result of the tension experienced by the working masses, and he extended the relationship between the city and cinema to a psychological model rooted in a Marxist understanding of labor: the modern metropolis creates working masses who are never compensated adequately; consequently their need for compensation can "be articulated only in terms of the same surface sphere that imposed the lack in the first place" (325, emphasis in original). Kracauer was critical of the surface quality of films, but he also argued that they revealed the living conditions of the masses in the modern metropolis.

Simmel, Benjamin, and Kracauer understood modernity in a Marxist context and thus analyzed cinema and city as part of the reorganization of labor and the market. Max Weber, on the other hand, focused specifically on the market as a rational and secular mechanism in the modern West. He argued that instead of relying on myth or religious beliefs, individuals in modern Western society make rational choices based on the calculations necessitated by capitalism. *The Protestant Ethic and the Spirit of Capitalism* (1905) contrasts the rational capitalist with the figure of the adventurer capitalist, who Weber assigned to the premodern, marked as geographically and temporally different civilizations: "Whenever money finances of

Figure 1.3 Advertisement for Walter Ruttmann's *Berlin: Symphony of a Great City* (1927)

public bodies have existed, money-lenders have appeared, as in Babylon, Hellas, India, China, Rome . . . This kind of entrepreneur, the capitalistic adventurer, has existed everywhere" (xxxiv). Instead of adventure capitalism, Weber emphasized "sober bourgeois capitalism" that relied on "technical possibilities" and "calculability" (xxxvii). *Metropolis* stages the encounter between adventure capitalism and the rational capitalism of the West that both supersedes and integrates religion, similar to Weber's analysis of Protestanism as an integral part of the development of Western capitalism.

The case study later in this chapter will discuss in more detail modern capitalism in relation to the modern city as understood in the early twentieth century. Before we get to it, however, two sections, one on the emerging genre of the street film and one on *Berlin: Symphony of a Great City* as a prime example of modernist aesthetics, reflect the theories outlined so far.

The Weimar street film

A subgenre of the city film that developed in the Weimar Republic was the street film, organized around the street as a space of random encounters, violent crimes, urban surveillance, and ambiguous morality and sexuality – the emerging social space and public sphere of modern urbanism. The fascination with the street reflected technological changes that enabled a new and different kind of street life in the city than previously existed. Frances Guerin explains that by "the 1920s, the industrialization of light that had begun in the mid-nineteenth century reached a moment of intensity" (155). Life in the streets became visible at night, which opened another dimension for interaction.

The street in the Weimar Republic street film became the setting for "psychological melodramas" that represented "a dangerous lure and a force of tragic destiny for the imprudent male" (Katz 520), the danger lying in the possibility of illegitimate desires bridging social divisions and moral codes. The inner workings of modernity were externalized in the urban modernity of the street, as we can see in Pabst's *Joyless Street*, which focuses on the working poor in its portrayal of class conflict.

Motherless working girl Greta, played by Greta Garbo, lives with her retired father, who loses his entire pension in an investment scheme of stock options for a mining company. A second plot involves Maria (Asta Nielsen), who is treated badly by her father and who in the course of the film prostitutes herself for her boyfriend, Egon, just to witness him with another prostitute, who she then kills. A group of ruthless adventure capitalists plays on the masses by pretending that coal-mine stock will go up, so that poor people invest and lose their money. The space most closely associated with the ruthless adventure capitalists is Madame Gill's Bar, a place of postwar decadence that provides a playground for them and represents the threat of prostitution for Greta and Maria as well.[3] The street is the space of social encounters across class.

Figure 1.4 G. W. Pabst. *Joyless Street* (1925)

Joe May's *Asphalt* (1929) portrays the dangers as well as the possibilities of those random encounters in the streets of the city. The film exposes the dangers of urban modernity by juxtaposing the transgression of law with the reconstitution of law, poles embodied by a prostitute and a policeman. An early scene shows the Potsdamer Platz with a policeman who is trying in vain to control the traffic with his hand stretched out, which positions him at the mercy of the modern metropolis. A low camera angle foregrounds the asphalt and the cars exceeding his control across which the title of the film, *Asphalt*, is written. The film constructs a stark contrast between interior domesticity, inhabited by the policeman's parents and coded as premodern, and the exterior urbanity of the modern metropolis, inhabited by the prostitute who seduces the policeman.

Like the prostitution and adventure capitalism in the space of Madame Gill's Bar in *Joyless Street*, the character of the prostitute in *Asphalt* shows the seduction of modernity and capitalism gone awry. Her cosmopolitan modernity connects her to crime across national borders in the form of a boyfriend who robs a bank in Paris. When the policeman visits her and her criminal boyfriend arrives, the

Figure 1.5 The shooting of *Joyless Street*

policeman kills the boyfriend out of jealousy and is consequently arrested. The figure of the prostitute is an early incarnation of the femme fatale of film noir in that she seduces a law-abiding man to kill her lover – see the readings of Billy Wilder's *Double Indemnity* (1944) and Howard Hawks's *The Big Sleep* (1946) in Chapter 2. The city film of the Weimar Republic was more optimistic about the possibilities enabled by these new social spaces than was the later cynical film noir, and *Asphalt*'s narrative has a happy ending – the prostitute testifies that the policeman acted in self-defense and promises to wait for him until he is released from jail. He has succumbed to the sexual seduction of the cosmopolitan metropolis – his fate was foreshadowed by his inability to control the traffic early in the film – but the prostitute is domesticated by his love, honesty, and morality.

In general, scholars disagree on how to read the gendered politics of the street film. At one end of the spectrum, Bruce Murray argues that, beginning with *Joyless Street*, the street films "promoted the maintenance of patriarchy" *vis-à-vis* the mysterious woman who threatened to undermine it. At the other end, Patrice Petro sees the street as the setting of "male symbolic defeat" (1989: 163). This binary underlies the femme fatale in the Weimar Republic city film and film noir as both castrating and empowering, and as a punishable and often punished character.

Whereas the films mentioned thus far project the seduction of modernity onto the woman-in-the street, Lang's *M* portrays "the darker side of the urban flâneur"

Figure 1.6 The empty street as setting in Fritz Lang's *M* (1931)

(Elsaesser 2000a: 145). As we have seen, in *M* a child murderer is terrorizing the city, and when the police efforts to catch him begin to impede the illegal activities of the underworld, members of the latter decide to capture the murderer themselves by mobilizing the beggars of Berlin. A blind street-vendor recognizes the murderer by his whistle – a brilliant use of sound in early sound films – and where modern police methods of surveillance have failed, the gangsters succeed in capturing the murderer using their knowledge and organization of those who inhabit the city. Anton Kaes suggests that *M*'s "obsession with surveillance also addresses the deep-seated fear of an expanding urban population," explaining that "Berlin more than doubled in population by the end of the decade; it had reached 4.5 million inhabitants in 1930" (2000: 49). *M*'s "conflict between surveillance and obscurity" connects cinema and the metropolis, according to Carsten Strathausen (25).

In *M*, the murderer's sadistic sexuality is presented through the narrative and mirrored in the social space of the city, but is evident also in the commodified space of a window display. At one point we see Beckert, the killer of little girls, looking into a shop window, his face framed by the reflection of knives displayed there in an ornamental pattern. The next shot is from Beckert's point of view as he looks into a mirror in the same window and sees a little girl outside, this time framed by the knives. Kaes reads this sequence of shots as "Beckert's desire . . . figured as violence" (2000: 60). The knives remain the consistent content across

Figure 1.7 The shadow of the mass murderer meeting his next victim

several edits, becoming a miniature mass ornament symbolizing that the commodity itself functions to transfer the violence to the object of desire. Kracauer developed the theoretical concept of "the mass ornament" (75), for which the point of departure was the Tiller Girls, a review group of young women who in their performances no longer appeared as individual performers but as "indissoluble girl clusters whose movements are demonstrations of mathematics" (76). The mass ornament resulted from a capitalist production process that destroyed natural organisms, community, and personality in order to create calculable entities. As Thomas Elsaesser points out, Kracauer's theory of the mass ornament situated the "relationship of self to body in terms of vision and self-display" (2000a: 48). The visual emphasis in the process of modernization affected not only cultural representation, but also self-perception and self-representation, a claim that is visible in the single shot of Beckert looking in the shop window.

In this short sequence the film accomplishes two things: one, it depicts Beckert as a passive character despite the aggressive violence of his crime, a facet which is acted brilliantly by Peter Lorre at the end of the film when he confesses but also cries out that he cannot help himself; and two, even though Beckert cannot control his destructive urges, the scenes of him gazing in shop windows show a reflexivity regarding representation and commodification that other cinematic texts lack in depicting women as commodities in the street. Thus, the sexual perpetrator is accorded both self-reflexivity and victim status by Lang.

Characteristic of the subgenre of the street film, then, is the street as the site of social interaction and control. It is also the space in which desires and anxieties are acted out. In accordance with the theories advanced by Benjamin, Simmel, Kracauer, and Weber, the street film is highly gendered: female figures appear repeatedly as prostitutes, and male characters range from enforcers of the law, in the form of policemen, to those who break the law, in the form of criminals. The gendered interactions are negotiated in the public space of the urban street.

Modernist aesthetics: the city symphony

While the street film offers melodramatic narratives acted out by characters, other films such as Ruttmann's *Berlin: Symphony of a Great City* reproduce the psychic and visual experience of modernity without relying on a conventional narrative. Hake accords *Berlin: Symphony of a Great City* a radical position because it organizes Berlin's "social and spatial qualities in visual terms" (1994: 127). However, despite the fact that many have applauded the film's aesthetic expression of the *avant-garde* and modernism, Hake maintains that Ruttmann's "ultimate goal was visual pleasure, not critical analysis" (1994: 127).

In Ruttmann's film, the modern metropolis of Berlin is anthropomorphized through the temporal organization of a full day there, from beginning to end, arranged in five acts. Onto the opening shot of calm waves of water and the sun, abstract shapes are projected: a circle, lines, and a square move in an abstract formation, accompanied by a short, atonal score. The abstract opening then cuts to a shot from a train moving towards Berlin through the surrounding gardens, industrial areas, construction sites, empty train stations, and advertisements, to the sign announcing Berlin. The very next sequence contrasts a close-up of a machine, signifying the anonymity and efficiency of modern production, with the old splendor of Berlin's cathedral shot from above. Human beings in the metropolis are continually subordinated to the material dimension of modernity in shots of modernist architecture, industrial design, and electricity. The emptiness of the streets in act one, at five o'clock in the morning, emphasizes both the absence of humans and the city as an entity in itself as the particular focus of the film.

The film's abstract opening of different shapes edited together and forming a rhythmic pattern is followed by a train-ride into the city. This sequence is not only a formal consideration, but reflects the fact that, as Kaes points out, "Berlin has always been a city of migrants from rural areas" (1998: 185). The film's opening shows nature and abstraction as two poles framing the idea of the modernist metropolis. Hake describes this model of editing as "a kind of associative montage," which, in contrast to the political commentary associated with Sergei Eisenstein, "confirms total exchangeability and eternal recurrence as the foundation of experience in modern mass culture" (1994: 130). She also points out that the film does not emphasize the monuments of the nineteenth century that identify imperial Berlin, an important change from the previous understanding of the city (1994: 134). The modern metropolis is marked mainly by the camera's repeated return to places in the city that are not identified by their national significance in the capital of Germany, but rather by their role for transportation or leisure. Instead of architecture, we find traffic, which continues the fascination with movement and brings into play the dynamic possibilities of traffic with the dynamic possibilities of editing. Ruttmann was a painter by training, and Strathausen explains that he published a statement before the film's release that characterized his montage technique as fulfilling "musical–rhythmic demands" (43).

Berlin: Symphony of a Great City repeatedly shows displays of mannequins in shop windows, thus combining their artificial, anthropomorphic quality with the seductive allure of commodities. The careful arrangements of mannequins to arouse the pleasure of looking and then seducing the implied *flâneur* to enter the store and purchase something are juxtaposed with the masses of people on their way to work, arriving for work, moving quickly and disconnectedly up and down stairs at subway stations. The shots are organized according to abstract principles

Figure 1.8 Walter Ruttmann. *Berlin: Symphony of a Great City* (1927): Traffic

of movement and composition, cross-cutting masses on the way to work with soldiers, marching in formation, and animals, and increasing the profusion of people from one shot to the next. Views of masses moving through the city, in turn, are intercut with close-ups of industrial, mechanical, and electrical machines that dwarf humans, and we also see instruments of communication, such as the typewriter and the telephone, which signal modernity.

Because there are no individualized characters, the few singular individuals take on a symbolic function, and once again one of the few is a prostitute, a streetwalker recognizable through her interaction with a man who picks her up. She is seen through the corner of a shop window, aligning her with both the seductiveness and sexualization of consumption and the public space of the street. The next contrasting shots show mechanical window displays and a wedding couple, pointing to the mechanization of sexuality in contrast to traditional matrimony. Frequent shots of neon signs dominate the cityscape announcing movies and reviews. Thus Ruttmann's film reflects aesthetically the experience of modernity, characterizing the city as Kracauer's "surface."

The street film captures the experience of modernity in narratives about urban types and projects the changing gender roles onto the newly emerging urban space of the street. *Berlin: Symphony of a Great City* responds to the experience of modernity as fragmented and abstract through its aesthetic choices of editing, rhythm, and rejection of traditional narrative. Both examples, however, reflect aspects of modernity highlighted by important theorists during the Weimar Republic, such as that of the *flâneur* and the metropolitan type, the configuration of the mass ornament, and the different formulations of the crowd; in short those kinds of configurations of cultural productions, urban spaces, and human subjectivity that changed with modernity but that also produced modernity in the city.

Case Study 1 **Fritz Lang's *Metropolis* (1927)**

Lang's paradigmatic city film *Metropolis* addressed the connection between the city and industrialization and served as a blueprint for science-fiction films by advancing film technology in order to present a vision of the urban future (see Chapter 6). The

Figure 1.9 Creating the set for Fritz Lang's *Metropolis* (1927)

city of Metropolis connects technology and futurism, socio-political problems, and their resolution. Anne Leblans sees in the film "a seismograph that with great accuracy registers concerns, conflicts, and developments of the mid-1920s" (96). Famous for its settings, the original production cost more than 4 million Reichsmarks, had a duration of seven hours, and did not recover its production costs.

As already mentioned, the imaginary city of Metropolis is organized vertically. The spaces above, pleasure gardens and the large office of the owner of Metropolis, Frederson, are inhabited by the upper class, while below the workers toil in uniform and drab workplaces and houses, dressed uniformly and moving in unison from and to work. Maria, the daughter of a worker, mediates between the workers and Frederson, who wants the workers to be replaced by machines to enhance productivity, but Rotwang, a mad scientist, builds a seductive, evil robot in the likeness of Maria to incite revolution among the workers and ultimately destroy Metropolis. Frederson's son, Freder, and the real Maria meet and fall in love. The robot Maria incites the workers to revolt, but their chaotic actions result only in the destruction and flooding of their own quarters. When their children are at risk of drowning in the flood, Maria and Freder save them. The foreman reports to the hysterical masses that their children had almost drowned, and consequently they burn the false Maria at the stake. Freder, Maria, and Rotwang fight on top of the cathedral, and Rotwang falls to his death. The final image of the film shows Freder, Frederson, and the foreman shaking hands in front of the cathedral with Maria at their side, and the last intertitle reads: "The heart connects the mind and the hand."

WIth Maria's help, Freder has come to be the connection between the alienated labor force and the owners of the means of production in industrial capitalism. Religious imagery is substituted for a political solution to the problem of class exploitation. Modern Weberian capitalism is embodied by Frederson, the boss of Metropolis, in his rational, profit-oriented relationship with those around him, including his son, Freder. Rational functionalism and automatism are also represented by many fetishizing shots of the machines that celebrate technology. Rotwang represents the danger of irrationality in capitalism, which could destroy the means of production as well as the workers. Ultimately, however, it is the spiritual connection of the heart between the hand and the brain, and the destruction of the atavistic and mystical–magical embodiment of capitalism that makes possible a modern, rational, and humane capitalism for the future. The film's narrative drives towards that final constellation encapsulated in its ultimate shot.

Metropolis opens with a shot of a city signaling its abstract design, which reflects the modernist vision of functionalism and the absence of decoration. Onto this shot

continued

are projected close-ups of industrial machines that function in a well-orchestrated rhythm. By showing only parts of machines, the close-ups celebrate the industrial aesthetic and at the same time obscure the concrete function of the machines. The next shot shows a ten-hour clock, symbolic of the organization of the industrial workday and the labor force in the shift-change of the workers. The workers are grouped as an anonymous mass and their alienation and exhaustion are expressed in their unison movement, uniform costume, lack of communication and absence of individualized facial expressions.

Figure 1.10 Mies van der Rohe's modernist vision of architecture

This lower class is contrasted with the playful and sexualized upper class in a space called the "Eternal Gardens" that is located above the workers' homes. Here the characters express their individuality through elaborate costumes and playful interaction. At the same time, their individuality is coded as superficial, particularly for the female characters. While modern in their self-confident expression of sexuality and extravagant clothing, they are also portrayed as excessively decadent, a quality mirrored in the design and architecture of "Eternal Gardens," with its fountain and peacocks reminiscent of baroque architecture, a stark contrast to the modernist design of city below. *Metropolis* captures the relationship of modernity in the city that is also the workplace through the scale of the relationship of setting and characters. Repeatedly the size of the architecture overshadows the workers, whose monotonous and repetitive movement also reduces them to parts of the machinery.

Several shots emphasize the city's fantastic modernist architecture with the pathways of the traffic below. While the film shows the conditions of work in the industrial city to be oppressive, it also celebrates the modernist design and architecture as aesthetic possibilities and, by implication, film's ability to showcase this splendor.

Figure 1.11 *Metropolis*: The spectacle of the modern cityscape

continued

Rotwang introduces the false Maria to upper-class men in a bar called Yoshiwara, a place where capitalist desire is sexualized. Yoshiwara is associated with the feminine and the Orient, echoing the notion of adventure capitalism described by Weber. This space – similar to Madame Gill's Bar in *Joyless Street* – connects the feminine in the commodification of sexuality with the atavistic form of capitalism. The threat of modernity is embodied by the figure of the prostitute, now also a robot, combining dangerous sexuality with the danger of technology. Andreas Huyssen points out that the figure of woman, split in "two traditional images of femininity – the virgin and the vamp, images which are both focused on sexuality, [poses] a threat to the male world of high technology, efficiency, and instrumental rationality" (1986: 72). The dialectic conclusion to the conflict is aided by characteristics that are rooted in the premodern.

Like so many other directors and films discussed here, Lang's *Metropolis* portrays the ambivalence of modernity. On the one hand, the modernist city gives birth to its own destruction embodied by the robot, representative of the technological displacement of humans. On the other hand, the impressive setting of the modernist city, with its oversized architecture, its celebration of machinery, and its reduction of humans to orderly masses, fetishizes the surface aspects of modernity, à la Kracauer. The film's lasting visual and intellectual appeal results from its reproduction and thus documentation of the fascination with an aesthetic vision of modernity more than from the critical discussion of modernity found in its narrative. Thus the fascination of audiences with the film also responds to the awe-inspiringly homogeneous, coherent, and larger-than-life vision of city that points to the possibility of a better future, a vision that is hardly visible today on the grand scale that Lang created in Neubabelsberg.

Further Reading

Thomas Elsaesser (2000b) *Metropolis*, London: British Film Institute. As part of the BFI series of film classics, Elsaesser's *Metropolis* advances a detailed historical account of important contexts of the film, such as Lang's own mythical account of the film's creation, the production at the studio, and the film's different versions.

David Frisby (2001) *Cityscapes of Modernity: Critical Explorations*, Cambridge: Polity. *Cityscapes of Modernity* provides an in-depth and detailed overview of the theories of modernity that this chapter can only address in a cursory fashion. Focusing on Berlin and Vienna, it pays particular attention to the relation between social theory and cultural production.

Anton Kaes (1996) "Sites of Desire: The Weimar Street Film," in D. Neumann (ed.) *Film Architecture: Set Designs from Metropolis to Blade Runner*, Munich: Prestel, 26–32. This essay analyzes the Weimar Republic street film as intimately tied to modernity.

Patrice Petro (1989) *Joyless Streets: Women and Melodramatic Representation in Weimar Germany*, Princeton, NJ: Princeton University Press. Petro offers a reading of the role of women in the Weimar Republic as spectators in the context of the theories of modernity and popular culture.

Jans B. Wager (1999) *Dangerous Dames: Women and Representation in the Weimar Street Film and Film Noir*, Athens: Ohio University Press. Wager's book connects the film noir femme fatale to her origin in the films of the Weimar Republic.

Janet Ward (2001b) *Weimar Surfaces: Urban Visual Culture in 1920s Germany*, Berkeley: University of California Press. Ward proposes that surfaces are an important concept for urban modernity as it emerged in 1920s Germany and elaborates on surfaces in architecture, advertisement, film, and shop windows.

Essential viewing

Fritz Lang. *Metropolis* (1927)

—— *M* (1931)

Joe May. *Asphalt* (1929)

F. W. Murnau. *The Last Laugh* (1924)

G. W. Pabst. *Joyless Street* (1925)

Walter Ruttmann. *Berlin: Symphony of a Great City* (1927)

Robert Siodmak, Edgar Ulmer, and Billy Wilder. *People On Sunday* (1929)

2 The dark city and film noir: Los Angeles

> You could charge L.A. as a co-conspirator in the crimes . . .
>
> Richard Schickel

Learning objectives

- To understand the definitions of film noir
- To position film noir in the urban post-Second World War context of the United States
- To discuss the aesthetic, topical, and biographical connections between Weimar cinema and film noir
- To decode the politics of gender in the representation of film noir's urban space

Introduction

This chapter continues the threads introduced in Chapter 1: the importance of cinema and cities for the modern world. The cycle of films addressed here, the black-and-white private-eye films made in 1940s Hollywood and set primarily in Los Angeles – called film noir by French critics – continued the aesthetic and topical features of the Weimar city film. Several of the important film noir directors were German or Austro-Hungarian and had fled Hitler's dictatorship in Europe; consequently, this chapter, and particularly the case study of Billy Wilder's *Double Indemnity* (1944), investigates the continuation of and breaks between the aesthetic and topical representation of the Weimar city film and film noir. Parallel to the first chapter, here I discuss Los Angeles as an allegory for modernity and also continue examining the ways in which gender functions in city films: the *flâneur* and the prostitute in the Weimar city film are reincarnated in the private detective and the femme fatale.

Defining film noir

Aesthetically film noir emphasizes the black-and-white qualities of light and shadow that create space and mood through low-key and chiaroscuro lighting. Skewed camera angles and lonely characters in empty, urban spaces evoke a sense of urban alienation. The city is usually shown at night and in the rain, and its inhabitants are morally corrupt and violent. Often voice-over narrations tell the story in flashback. Families are incomplete and characters betray and double-cross each other. The crisis of masculinity coincides with the presence of the femme fatale, a sexualized, double-crossing, dominant female character who is ultimately punished for her transgressions.

The Hollywood studio system also affected the production of film noir, especially the star system and the artistic division of labor according to which screenwriting, producing, and directing became separate functions undertaken by different people. Actors, directors, and screenwriters were employed for a certain number of films or years with a studio such as Paramount, Universal, MGM, Samuel Goldwyn, Twentieth-Century Fox, Warner Brothers, or United Artists. Because of the division of labor and because the studio's overall goal was to turn a profit, the studio system favored developing and reproducing genres, generic conventions, consistent cinematic styles, and film cycles.

Los Angeles as a setting for film noir reflected pragmatic production considerations but also imbued the real city of Los Angeles with a symbolic dimension of alienated urban space. As Edward Dimendberg has carefully outlined, Los Angeles in the 1950s shifted from a centripetal organization of urban space to a centrifugal organization and experience of urban space (see 86–118 and 166–206). Film noir preferred transitional spaces, such as hotels and train stations, streets and alleys.

Despite its wide and frequent use by academics, cinephiles, lay people, and those working in the film industry, and although it seems obvious which films can be so labeled, 'film noir' is highly contested in literature on the topic. While a certain number of films put out by Hollywood within a particular period that look a certain way are identified as film noir, there is disagreement about what kind of systematic category organizes these films: is film noir a genre, a cycle, or a style? I suggest that the rather academic controversy results from the birth of the term, in France, years after the first films were made, thus with a temporal and geographical gap.

During the Second World War, Europeans were unable to watch American films, and afterwards, when they saw the films made in Hollywood during the war, they were struck by the homogeneity of their style and content, and what they perceived as a darkness that sharply differentiated these from American films as they had known them. In 1955, French critics introduced and coined the term 'film noir,' which Raymond Borde and Étienne Chaumeton elucidated upon in the same year in their groundbreaking, book-length study *Panorama du film noir américain* (1941–1953) (*A Panorama of American Film Noir 1941–1953* [2002]), which was then reprinted with an expanded filmography including selected films from a more expansive time period, 1915–74 (165–228).[4] Following the time-lag from the French invention of the term and the American acceptance of and recourse to that definition, American writers in the 1970s began to discuss it in a "cross-referenced series of essays" (Silver and Ursini 1996: 3), and Hollywood used the term "as a marketing tag for films earlier labeled as 'melodramas,' 'thrillers,' or even 'psychological chiller-dillers'" (Dimendberg 5). A steady stream of publications on film noir throughout the 1980s ultimately led to an explosion of academic and popular writing on the topic in the United States in the late 1990s.

In 1983, Foster Hirsch defined film noir with reference to two representative films associated with two representative cities: *Double Indemnity*, which takes place in Los Angeles, and Fritz Lang's *Scarlet Street* (1945), which takes place in New York City. Hirsch claimed that these two films are "as representative of the genre as *Stagecoach* is of Western or *Singing in the Rain* of musicals" (2). The assumption that film noir is a genre as clearly defined as the two other paradigmatic American genres, the Western and the musical, has been and is still contested among those who write on film noir or genre film. In 1995, Paul Schrader claimed that film noir is not a genre, but, ironically, his essay was republished in the collection *Film Genre Reader II* – a symptom of the proximity of the conflicting opinions. In other words, even those who claim that film noir does not constitute a genre have to acknowledge that it looks much like and therefore is much like a genre. Schrader explains: "It is not defined, as are the Western and gangster genres, by conventions of setting and conflict but rather by the more subtle qualities of tone and mood" (214).

Hence Alain Silver in his 1996 *Film Noir Reader* prefers the term "noir cycle," because film noir is set apart from other Hollywood films by "the unity of its formal vision" (4). Dimendberg agrees, following Borde and Chaumeton, who defined it as "a historically circumscribed group of films sharing common industrial practices, stylistic features, narrative consistencies, and spatial representation" (11). Importantly, for Dimendberg it is the conception of space that cannot be captured in the definition of genre or style: "the film noir cycle reveals practices of representing and inhabiting space and suggests how culture itself can be understood as a mode of representational and spatial practices" (11). In conclusion, following Borde and Chaumeton, Silver, and Dimendberg, I employ the term "cycle" as the basis of the discussion of film noir here, because this chapter, as a microcosm of the book's overarching theme, is concerned primarily with the relationship between modes of production, formal features, and urban spatiality.

The urban spaces of film noir

Film noir associates the city with alienation, isolation, danger, moral decay, and a suppressed but very present sexuality. The alienation of characters finds expression in their repeated movement alone through the urban space and their chance encounters with other lonely characters. One striking example is found in the opening of Robert Aldrich's *Kiss Me Deadly* (1955), and it motivates the film's narrative. The film begins with a full minute showing a screaming woman running up an unidentified street towards the camera. A man in a sports car picks her up. Soon thereafter, the police stop them, looking for an escapee from an asylum. The strangers pretend to be a married couple, turning bourgeois marriage into a charade employed to pass the law. A second sudden encounter, this time with gangsters, leads to the torture and murder of the mystery woman and the man being left for dead at the side of the road. The narrative develops from these opening chance encounters, when it is revealed that the main character is detective Mike Hammer, who then searches for the secret behind the mysterious encounter on the highway into Los Angeles.

Film noir associates the city also with a lack of emotion, and an acting style developed according to which actors delivered their lines solely by moving their lips. The link between the city, alienation, and emotional detachment continues Georg Simmel's notion of the metropolitan type resulting from "the intensification of nervous stimulation" (175). The typical character in film noir reflects Simmel's description of the "blasé attitude," a psychic phenomenon "unconditionally reserved to the metropolis" (178). In his explication of that type, Simmel followed a turn-of-the-century model of "nervous stimulation" that could lead to nervous collapse and was associated with feminization: "A life in boundless pursuit of pleasure makes one blasé because it agitates the nerves to their strongest reactivity for such a long time that they finally cease to react at all" (178). It is therefore not surprising that often femmes fatales personify the blasé attitude. In Howard Hawks's *The Big Sleep* (1946), for example, the two daughters Vivian and Carmen Sternwood display the blasé attitudes of those who have indulged in so much pleasure that they have become simultaneously immune and addicted to it, particularly in relation to the pain and murder that goes on around them.

The Big Sleep subtly differentiates between generations of Los Angelinos and associates them with different urban settings, leading Philip Marlowe, the hero, through a set of urban spaces from the library to gambling casinos. But the film begins in a setting that represents Los Angeles more symbolically. In the opening sequence, private detective Marlowe visits General Sternwood, who lives with his two grown daughters in a mansion overlooking the oilfields that represent old, corrupt Los Angeles.[5] Sternwood does not embody the blasé attitude of the urbanite and is concerned about his two daughters. The scene opens in Sternwood's humid, overheated greenhouse where he grows orchids – a metaphor for his daughters – and where the old man is indulging in the vicarious pleasure of having his guests drink and smoke. Soon enough we – the audience – along with Philip Marlowe encounter the daughters, the next generation, raised on old money in a new city, and embodying the blasé attitude. The orchids symbolize the overcivilization associated with urbanization mirrored in the two upper-class daughters' addiction to drugs and gambling.

Streets and alleys, shown primarily at night and in the rain, are the more obvious urban settings of film noir. They provide the environment for alienated characters, chance encounters, and chases, which in turn motivate narratives of mystery and detection. Interior spaces often portray the seedy but extravagant underworld of the city. Thus, singers and band leaders appear in lavish casinos and bars, providing an aural context associated with postwar urbanity, often performed by African-Americans and rooted in jazz. Consider what Nicholas Christopher describes as "one of the most expressionistic of all films noirs," Robert Siodmak's *Phantom Lady* (1944), which is however set in New York City (203). Scott Henderson is suspected of murdering his wife, and his secretary, nicknamed Kansas, attempts to prove his innocence. To find the real killer she searches for a mystery woman who was Scott's date after they met in a bar and, without exchanging names, went to hear the singer Estela Monteiro. The only characteristic identifying the unknown woman was her unusual hat, which was the same as the one Estela Monteiro wore during her performance.

In order to track the mystery woman, Kansas attends the Broadway show and seduces the drummer of Estela Monteiro's band, who takes her to a jam session in an alley basement where jazz musicians are playing in a scene that Christopher calls "one of the steamiest simulations of sexual intercourse in film, with neither participant ever once touching the other" (206). The more the drummer plays his drums, the more Kansas reacts with her body. During his solo the camera moves back and forth between his sweaty face and her passionate face until their "climax, [leaving] them both looking exhausted, post-coital" (206). Sexual activity is associated with and expressed through sound, because it cannot be shown on the screen. The film's narrative is mapped onto the topography of the city, splitting it into the visible, formal performance space of Broadway and a hidden, informal space characterized by unscripted performance and improvised jazz, which in turn references African-American urban culture.

Christopher sees this particular scene as an example of the film studio's attempt to circumvent the Hays Office censors. From 1934 until the mid-1950s, the content of films was controlled by the Hays Office Production Code, which included "Particular Applications" for the depiction of crime and sex, demanding primarily that sympathy should never be created for a crime, that killings should not be presented in detail, and that revenge should not be justified (Belton 139). Showing illegal drug-trafficking and the consumption of liquor was not allowed even for the sake of the plot or characterization. Adultery could not be shown as attractive, and excessive kissing, lustful embraces, suggestive postures and gestures, were all prohibited, as were "[s]ex perversion" and "[m]iscegenation."

The important settings of film noir where illegal activity and sexualized encounters could take place had to be integrated into the narrative and their attractiveness disavowed by it. The gambling casino and the urban bar function as sites for narrative turning points, revelations, and clues in detective stories. The bar is the urban space, above all others, where unattached members of the city meet, exchange with each other, and then go their separate ways, their encounters often crossing class, gender, and racial lines. In *Phantom Lady* the chance encounter in the bar proves to be the key to exonerating the main character, Scott Henderson. To prove his innocence, Kansas has to return there, a woman alone in a space traditionally identified as male. Similarly, in *The Big Sleep*, Vivian Sternwood is singing to the jazz piano in Mr Mars's gambling joint when Philip Marlowe enters, and in *Kiss Me Deadly* an African-American woman is singing Nat King Cole's *Rather Have the Blues* in a club called "Pigalle." The figure of the black bartender appears repeatedly in film noir as the signifier of repressed race, a subtext of film noir's anxiety about urban space.

Whereas bars, casinos, and hotel lobbies are sites of anonymous encounters and immoral behavior, and are important in delineating characters, the hotel room is equivalent to the domestic space of the home and is thus another favorite transitional space in film noir. For example in Fritz Lang's *The Big Heat* (1953), Det. Sgt David Benyon lives in a hotel room after his wife is killed, and Debbie Marsh, the former girlfriend of a gangster, visits him there at night. The time she spends with David Benyon at the hotel is emblematic of the usual relationship in the world of film noir: isolated, transitional, and inhabiting spaces that are neither bourgeois nor domestic.

Since film noir is associated primarily with urban spaces, the narratives seldom venture into suburbia; however, in the late 1940s and early 1950s Los Angeles began to move towards suburbanization and decentralization – changing the "bodily experience" and "valoriz[ing] speed" (Dimendberg 169). As a result, in the late 1940s urbanity lost "its former monopoly as the dominant spatial mode of the film [noir] cycle" (211), and southern California became the model for "centrifugal identity" (168) as portrayed in important examples of film noir in 1949: Jules Dassin's *Thieves' Highway* (1949), Robert Siodmak's *Criss Cross* (1949), and Raoul Walsh's *White Heat* (1949) (see Dimendberg 177).

The Big Heat contrasts the suburban home of detective David Benyon with the urban underworld of Los Angeles: danger emerges from the city and encroaches violently on suburbia, destroying the detective's family and home. At first the film shows Benyon extensively in his suburban home with his little daughter and wife. He helps his wife with housework and does the dishes, while she shares his cigarette, scotch, and beer. At dinner they exchange sexual innuendos, and when they kiss, his wedding ring is in the center of the screen. Domesticity in suburbia and the identity of an urban detective are then shown to be incompatible when his wife is killed by criminals. The film repeatedly crosscuts suburban domestic life with bar scenes, contrasting legitimate with illegitimate sexual desire and legal with illegal labor, but, as the narrative continues, it moves increasingly into the city. The film acknowledges the space of suburbia, but narrative and aesthetics ultimately opt for the danger, violence, destruction, and immorality of the city.

Los Angeles as setting

Even though other famous American cities – New York City, San Francisco, Chicago – feature in individual films noir, the setting of Los Angeles is central to the cycle because of its position between modernity and postmodernity. Los Angeles appears as an imaginary city without history, or haunted by its history as in *The Big Sleep*. Dimendberg proposes that "the Californian subset of film noir is but a key chapter of its more expansive treatment of late modernity, a process of reconfiguring place and identity that exceeds a single geographic locus" (19).

Typical of film noir is the establishing, identifiable shot or sequence of shots of Los Angeles, usually at or near the beginning of the narrative. City Hall may be identified through a voice-over or by the characters, or there may be a shot of an identifying sign, such as the plaque that reads "Hollywood Public Library" introducing the second setting of *The Big Sleep*. The opening of *Criss Cross* shows us the Los Angeles skyline before the camera pans down into a parking lot in front of a bar and stops in a medium close-up midway through an intimate conversation between a man and a woman. From there develops a story of betrayal, murder, double-crossing, victimized men, and a femme fatale that is also the story of a return to the seductive and destructive facets of Los Angeles. An ironic variation of the establishing shot can be found in Billy Wilder's *Sunset Boulevard* (1950), which begins with the name of the street "Sunset Boulevard" and a tracking shot showing us just the asphalt of the street: the famous Los Angeles street is named, but the glamor associated with it is completely withheld.

Certain buildings – like Union Station, a recognizable, impressive, modernist building – reappear in several films, such as *Criss Cross*, Russell Rouse's *D.O.A.* (1950), Anthony Mann's *T-Men* (1947), and the neo-noirs *Hickey & Boggs* (Robert Culp, 1972) and *Heat* (Michael Mann, 1995; see Silver and Ursini 2005: 98 and 99). Los Angeles is central to film noir in two senses: on the one hand, films emphasize the postmodern quality of Los Angeles to portray alienation and rootlessness; on the other hand, noir narratives and aesthetics also create Los Angeles as an alienated, transitory, postmodern city.

Film noir and gender

French theorists after the Second World War were struck by the extreme gendering of film noir, and many feminist critics, particularly in film theory, continue to emphasize film noir's staging of the crisis of masculinity, the femme fatale, and the incomplete family (see Krutnik; Kaplan). Much of this discussion relies on psychoanalysis as a methodology, in part because the films themselves popularized Freudian psychoanalysis. Frank Krutnik sees the psychology of crime as an indicator of the increased importance of psychoanalysis in film noir, noting that interest in criminal psychology, the motive, and the psychological effects of a crime led to the narratological conventions of the voice-over and flashback. The RKO B-film *The Stranger on the Third Floor* (Boris Ingster, 1941), one of the first films noir of the 1940s, includes "several flashbacks, voice-over narration, and an extended dream sequence which was seen at the time as influenced by the German Expressionist film *The Cabinet of Dr Caligari* (1919)" (Krutnik 47). Several of the directors of film noir had fled Hitler, including Fritz Lang, Billy Wilder, Robert Siodmak, Otto Preminger, Curtis Bernhardt, and Edgar Ulmer. Krutnik points out that the film's dream sequence relies on a "distorted *mise-en-scène*" that symbolizes "the hero's psychological destabilization" (49). According to Krutnik, film noir also substitutes psychological innuendo for explicit sexual representations because of the Hays Code restrictions, which resulted in linking "Freudian psychoanalysis and hidden or illicit sexuality" in the mind of the public (50).

While Krutnik focuses on the crisis of masculinity, many feminist discussions about gender and film noir center on the femme fatale (see Doane; Wager; and Kaplan). Mary Ann Doane traces the femme fatale back to the figure's roots in the nineteenth century. The femme fatale is never "what she seems to be," she has a "secret, something which must be aggressively revealed, unmasked, discovered," and she is therefore often killed as punishment for her sexual and moral transgressions (1). Doane interprets this as a "reassertion of control on the part of the threatened male subject" and views the femme fatale not as a "heroine of modernity," but as "a symptom of male fears about feminism" (2–3).

In contrast to Krutnik and Doane, Jans B. Wager shifts the discussion of the femme fatale from the crisis of masculinity to the question of spectatorship. Following Patrice Petro's argument about the female spectator of Weimar cinema, Wager argues that both Weimar street film and film noir "assume and address a female spectator" (xv). Wager suggests that the femme fatale in both periods offered an independent pleasure to female spectators as a transgressive character who is recontained in the last reel. She traces the connection between the Weimar street film and film noir through the femme fatale via the urban environment and the historical context of postwar conditions – be it the First or the Second World War – when public opinion turned against working women.

The femme fatale is associated with aesthetic choices that differ from Hollywood's traditional conventions for photographing female characters, such as soft-focus or diffused lighting. In contrast, "noir heroines were shot in tough, unromantic close-ups of direct, undiffused light which create a hard, statuesque surface beauty that seems more seductive but less attainable, at once alluring and impenetrable" (Place and Peterson, 328). Wager connects these aesthetic choices to the urban setting, which "create[s] an atmosphere of uncertainty and fear; and foreground[s] female sexuality and duplicity" (78), in contrast to the domestic realm and the trapped woman, or *femme attrapée*, who sacrifices herself for the man (15). Wager emphasizes the gendering of the two spaces that organize film noir: "The urban realm, at once the province of danger and desire, contrasts with the quotidian existence represented by the wife or mother and the domestic realm" (118).

In *Criss Cross* the femme fatale and the city stand in for each other. The more the main character, Steve, who returns to Los Angeles, is seduced by his former love Anna, the more he is enveloped by the city. When he arrives back in Los Angeles, the narrative is moved forward by his search for Anna, which he does not acknowledge and which therefore characterizes him as unaware of his own motivations. In order to find Anna, he has to venture repeatedly and ever deeper into the city, and a scene at Union Station portrays the parallel function of the woman and the city. Steve's chance to escape the lure of both is "criss-crossed" at the station, when he intends to leave but runs into Anna and consequently stays in Los Angeles. In the voice-over he articulates the collapse of the city and the woman into one another: "Everywhere you go you see her face." The femme fatale and the city are mutually constitutive, reinforce one another, and can stand in the one for the other: both are shiny, surface creatures without moral depth, who seduce men into their hold. Ultimately, both are unreliable and destructive. While this image of femininity is a particularly coherent version of older tropes that can be traced back via the Weimar city film to nineteenth-century conceptions of woman, the conflation of the femme fatale and the city is characteristic of the postwar depiction of Los Angeles in film noir.

Traces of Weimar cinema

Discussions of film noir aesthetics link the cycle to Expressionist cinema and interpret the presence of so many European immigrants in Hollywood as proof of that continuity. However, if we look closely, most of the directors of film noir in Hollywood who had made city films in the Weimar Republic – Robert Siodmak, Fritz Lang, Billy Wilder – had been indebted to New Objectivity and not Expressionism. John Willett defines the former as "objectivity in the sense of a neutral, sober, matter-of-fact approach, thus coming to embrace functionalism, utility, absence of decorative frills" (112), quite a contrast to the "passionate distortions of Expressionism" (11). But why did Expressionist aesthetics emerge in post-Second World War Hollywood among artists who had not participated in the Expressionist movement in Weimar Germany before they left?

I suggest that it is not a linear development from Weimar city film and Expressionism to the Hollywood studio system and film noir. Instead, directors who embraced New Objectivity for the realist depiction of the city in Weimar cinema moved to Expressionism to depict American urban spaces in film noir. Expressionism was the artistic response to the trauma of the First World War and emphasized the artist's inner vision, while New Objectivity's realism included optimism towards urbanity and technology. Expressionism distorts perspective, emphasizes gesture and types, and articulates an iconographic language for the subconscious. The Expressionist cinema of the Weimar Republic, exemplified by Robert Wiene's *The Cabinet of Dr Caligari* (1919), depicts a universe of unconscious desires and anxieties.

The traces of Expressionism in post-Second World War Hollywood are based on several interlocking factors. The repression of sexuality requires a cinematic language that translates the desires into a non-realist representational system; also, the cinematic language of Expressionism allowed German-speaking exiles to access and translate their own trauma and relationship to the alien city into a language of archetypes. Consider the relationship of the exile directors to Europe at the time. Their departure, for the most part in 1933, was initiated by the exclusion of Jewish and left-leaning directors in the film industry, and they were desperate to survive in Hollywood. Some of them, including Billy Wilder, embarked on successful careers at the very time when the true extent of the horror of the "Third Reich" was slowly revealing itself, creating a paradoxical situation for those directors. According to Ed Sikov, "Hitler led his troops into Vienna, annexed Austria in a flash of military might, and immediately supervised the systematic torture of Vienna Jews. With ghastly precision, the *Anschluss* occurred during the very week in which Billy ought to have been enjoying the release of his first major American film" (124). The increasing violence in Europe, particularly against Jews, amplified the sense of a living nightmare for those who participated in the machinery of entertainment, Hollywood.

Expressionism as a cinematic language is fed by the trauma of the past even when it is forward-looking (as in the left-wing politics of revolutionary drama, prose, and poetry). At a time of intense trauma in the lost homeland, directors employed a language that expressed their present through their past. Even though Wilder had been one of the most successful émigré screenwriters (though most often paired with Charles Brackett), the question of linguistic translation remains. Cinematic language, particularly if it finds a way to translate the unconscious, needs no translation. The Expressionism of the Weimar Republic remains a recognizable trace in film noir but is not entirely the same. The Expressionist style of film noir creates an alienating effect in conjunction with realist depiction of urban spaces; we never doubt that we see Los Angeles, for example. The setting captures realist representation; it is primarily the lighting that creates the Expressionist effect. Studio-driven film production mainstreamed an artistic vision into a cinematic, reproducible, and recognizable style. According to Sikov, Wilder saw himself as part of an industry and was not opposed to its economic underpinnings: he said, "We are an industry. There's nothing wrong with that, when you know you're commercial and aren't under any illusions of doing something else" (124). Wilder's success is based in large part on his ability to translate art into commercial viability.

Case Study 2 **Billy Wilder's *Double Indemnity* (1944)**

Double Indemnity opens with a car racing through the empty, nightime cityscape of Los Angeles. A man walks into an office and sits down at a dictaphone to record a story about a murder. Then the story is told through his voice-over and an accompanying flashback: Walter Neff, who provides the voice-over, is an insurance agent who visits Phyllis Dietrichson in a Spanish-style mansion in Los Angeles to renew her husband's insurance. Walter is seduced to return the next night when neither the husband nor the maid is at home. Communicating mainly by innuendo, Walter and Phyllis develop a plan to kill her husband and defraud the insurance company, which pays double if the insured is killed by a train. After Phyllis visits Walter at his apartment, he returns to the mansion to sell insurance to her husband, and they continue to meet secretly in a convenience store to finalize their murderous plan.

When her husband breaks a leg and has to walk on crutches, Phyllis takes him to the train station but shoots him on the way, and then she and Walter stage a deadly train accident. Walter's boss, Keyes, has a hunch that something is wrong, so Walter insists that he and Phyllis can no longer see each other. Meanwhile Lola, Dietrichson's daughter from his first marriage, visits Walter and tells him that Phyllis had been her mother's nurse and had played a role in her death. Walter's boss figures out what's going on, and Lola tells Walter that her former boyfriend, Nino Zachetti, is at Phyllis's house every night, thus exposing her as a double-crossing vixen who has used him. Finally Walter visits Phyllis, who shoots at him but is herself killed by him. The film ends with the return to the framing narrative of Walter's confession in the dictaphone to his boss Keyes and his subsequent collapse at the glass door of his office, where Keyes finds him when he arrives in the morning for work.

Expressionism characterizes the play of light and shadow throughout the film. During the credits we see the shadow of a man on crutches, a foreshadowing of the narrative, reminiscent of the expressionistic hand at the beginning of Fritz Lang's *M* (1931), but a realistic outline of a man. *Double Indemnity* uniformly emphasizes extreme light and shadow, always associated with distinct and carefully placed light sources: as Walter sits at his desk, the lamp creates a distinct light source, and additionally, he lights his cigarette. During daylight stark shadows are created even outside among palm trees. The interior space of the Dietrichson living room is characterized by stark shadows where the play of the sun through the blinds seems almost plastic, and the air appears heavy and suffocating. Phyllis is accompanied by extreme hard lighting that casts shadows when she walks through her living room, which looks like a museum, creating a disjuncture between the detached, sexual presence of the femme fatale and the space presumed to evoke domesticity. Walter's dark, bare apartment keeps him "in the dark" and is in extreme contrast to the California light outside. When Phyllis visits him, she stands next to the lamp, so one side of her face is brightly lit while the other remains in shadow, a symbol for her double personality. The film includes few of the skewed angles typical of film noir, relying mostly on straight-on shots except when a low-

continued

angle shot of the husband captures the suspense and danger on the way to the train-station. At the final meeting in Phyllis's house, Walter asks her to turn off the lights, so the setting for the final shooting is extreme darkness pierced by only a few light sources.

Double Indemnity created a legacy of stock characters for film noir. Phyllis is the archetype of an ice-cold, femme fatale who killed her husband's first wife and now uses Walter to kill her husband. (She also, it turns out, stole Lola's lover.) In contrast Walter, even though he becomes a murderer, is relatively passive and is seduced into crime. Richard Schickel summarizes the noir paradigm thus: "an all too compliant male is enthralled by a strong and scheming woman" (21). Phyllis lives in the overstuffed and dusty Spanish mansion, while Walter inhabits the lonely life of a bachelor. His only social connection in the film is his boss, Keyes, described by Claire Johnston as "the idealized father of Neff" (91). Even though Phyllis seems like a kept woman, she does not match the stereotype of the bourgeois housewife because she had worked as a nurse. Phyllis's dusty house with stuffy furniture appears outdated for her modern character. And while Walter matches the world of Los Angeles as a product of the alienation and loneliness of the city, it is precisely that alienation and loneliness that make him susceptible to the erotic promise of a woman.

In the final scene of the film, Phyllis confesses to Walter: "I never loved you. Not you, not anybody else," suggesting a lack of traditional femininity defined by tender love and care. The femme fatale manipulates the male character by pretending love and femininity, and his inability to recognize the charade is his ultimate downfall. The feminist attraction to the femme fatale is not a naïve celebration of her illusionary power, but is tied to the fact that she foregrounds the performance of femininity. Patriarchal order has to be re-established through Walter's destruction of Phyllis and confession to the patriarchal authority, his boss. Walter not only shoots Phyllis but also re-establishes the order of all the other relationships that she manipulated when he sends Nino Zachetti back to Lola.

Double Indemnity expresses Walter's alienation by having him tell his story to a dictaphone, but it also generalizes the alienation of the city. Schickel explains: "A bowling alley, a drive-in, a grocery store, the exterior of the Dietrichson house – there was something isolated, unwelcoming about all of them. Oh yes, Billy Wilder knew something in his bones about being rootless in Los Angeles" (61).

Wilder did little location shooting; instead, the characters reference Los Angeles repeatedly in their dialogues and map the narrative events onto the city. Walter's voice-over marks the precise location for each turn of the narrative. At the beginning, even before he names locations, those who know Los Angeles recognize that he is driving down Wilshire Boulevard past LaFayette Park. Later on in the film he secretly meets Phyllis at "the big market" in Los Feliz. Silver and Ursini describe the actual filming: "Billy Wilder directs Barbara Stanwyck and Fred MacMurray in Jerry's Market while police officers guard the food, an extremely valuable commodity during World War II rationing, used as set dressing" (2005: 28). When Phyllis's husband leaves Los Angeles by train, he is leaving from "Glendale," and characters repeatedly refer to Palo Alto, north of Los Angeles, particularly with regard to Mr Dietrichson being a "Stanford guy." After the murder, when Lola is suspicious of Phyllis, Walter tries to distract her and takes her "to dinner on Olivera Street" and on Sunday "on a ride to the beach." Even at the end, when Keyes calls for an ambulance, he emphasizes the location as Olive Street.

These geographical markers anchor the narrative of passion and deceit and the Expressionist aesthetic in the realist claim to the specific urban site of Los Angeles. The narrative events coincide with the topographical markers, suggesting a mutual dependence between the specific locale and its significance for viewers.

Further reading

Raymond Borde and Étienne Chaumeton (1955) *Panorama du film noir Américain*, Paris: Editions de Minuit; translated by Paul Hammond *A Panorama of American Film Noir, 1941–1953*, San Francisco, CA: City Lights (2002). The significance of Borde and Chaumeton's book lies not only in its survey of film noir, but also in the fact that it is seen as the historical moment when film noir was named.

Edward Dimendberg (2004) *Film Noir and the Spaces of Modernity*, Cambridge, MA: Harvard University Press. An extremely well-written book that discusses film noir in the context of trends in urban theory, while also paying close attention to the novels on which films noir are based.

E. Ann Kaplan (ed.) (1998) *Women in Film Noir*, London: British Film Institute. This volume gathers classic texts regarding the relationship of gender and film noir.

Alain Silver and James Ursini (eds) (1996) *Film Noir Reader*, New York: Limelight. This is the first of a series of readers on film noir and as such brings together canonical and foundational texts.

Essential viewing

Howard Hawks. *The Big Sleep* (1946)

John Huston. *The Maltese Falcon* (1941)

Robert Siodmak. *Phantom Lady* (1944)

Billy Wilder. *Double Indemnity* (1944)

—— *Sunset Boulevard* (1950)

3 The city of love: Paris

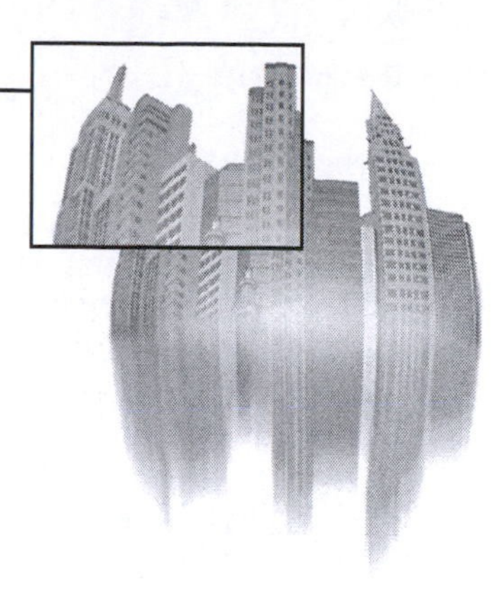

I believe that in cinema, there can only be love stories.

Jean-Luc Godard

Learning objectives

- To define *auteurism*
- To understand the French New Wave in the context of technological innovation, generational conflicts, and aesthetic concerns
- To advance a coherent reading of a film based on the depiction of urban spaces and the characters' movement through the city

Introduction

We now move transcontinentally to Paris and the French nouvelle vague (New Wave), a movement that explicitly articulated a theory of *auteurism*, a film movement contraposed to the studio system. The filmmakers who constituted the movement – François Truffaut, Jean-Luc Godard, Eric Rohmer, Agnès Varda,[6] Claude Chabrol, and Jacques Rivette – expressed their understanding of the nouvelle vague through the role of the *auteur*, the director who relies on improvisation in respect of the script and the acting. The nouvelle vague and Paris are linked to the degree that the relationship appears self-evident, and few scholars have investigated the conditions and implications of that connection beyond the fact that several of the important *auteurs* of the nouvelle vague grew up in Paris.[7]

This chapter attempts to fill this academic lacuna with an emphasis on the ways in which Paris and the nouvelle vague are constitutive of each other. The films of the

French New Wave favor on-location shooting, enabled by fast film stock that requires less light and by lightweight cameras, a small crew, direct sound, and amateur or lesser-known professional actors to create an experience of authenticity. Just as film noir reworked the Weimar city film, nouvelle vague cites film noir based on their shared concern for the staging of urban space, as the case study of Jean-Luc Godard's *Breathless* (1960) will demonstrate.

The Paris syndrome, or the city of love

In October 2006 a Reuters press release reported the "Paris syndrome," first diagnosed in the French medical journal *Nervure* in 2004, with symptoms of accelerated heartbeat, giddiness, shortness of breath, and hallucinations. The Paris syndrome apparently affects only the Japanese and only when visiting Paris; women in their early thirties on their first international trip are particularly prone to the affliction. The Japanese Embassy repatriates about 25 Paris visitors per year who claim to suffer from this condition named after a world city that once was the traveler's dream destination (see "Paris syndrome"). We can assume that the cognitive disjuncture and nervous breakdown result from a kind of culture shock at encountering the real Paris, not the Hollywood notion of romance that conventional feature films associate with the city. However, the notion of love associated with the French New Wave cinema also differs radically from the sugar-coated version of contemporary Hollywood.

The "city of love" seems a far cry from the dark, urban underside of film noir, but love in the nouvelle vague is not just the romantic love associated with Henri Cartier-Bresson's black-and-white photos or recent romantic projections onto Paris in contemporary Hollywood films. Love in the nouvelle vague has rough edges, is fraught with exploitation, disillusionment, betrayal, pain, suffering, and occasionally death. But love is also celebrated as beautiful through a set of aestheticized objects: film, Paris, and woman as object of desire. These (sometimes nostalgic) celebrations are articulated through an aestheticized realism made possible by new postwar technology, especially lightweight cameras. The tropes established in the Weimar city film and continued in film noir endure, especially that the association with the city connects love to lust, betrayal, and sometimes crime. The interweaving of love and the sensual and visually stunning portrayal of Paris evokes an affective response to the city and suggests a highly emotional relationship between the *auteurs* and Paris. The biographical dimension of several of the filmmakers who grew up in Paris – Truffaut, Godard, and Chabrol, for example – differs from the relationship of Weimar filmmakers with Berlin, most of whom had migrated to the metropolis from within Germany and the Austro-Hungarian Empire, or the film noir directors' relationship with Hollywood, of whom many were national and transnational migrants. Yet while I acknowledge this biographical dimension at the outset, the characteristic of intimacy in the nouvelle vague's portrayal of Paris exceeds a biographical explanation.

"No city, perhaps, symbolizes a nation as surely as Paris does France," posits Naomi Greene (247). Although Berlin was the capital of Germany during the Weimar Republic, in the city film Berlin signified primarily modernity; Los Angeles is not the capital of the United States, and in film noir it came to epitomize the postwar urban dissolution of social structures. Greene claims that Paris always had a privileged presence in French cinema, but emphasizes the New Wave's depiction "of post-war social dislocations and anxieties" (247). Specific changes in the city included "the gentrification of working-class neighborhoods; the growth of affluent suburbs as well as the rise of desolate housing projects; the influence of globalization and of the kind of consumerist culture associated with the USA." Historian Robert Gildea calls the period from 1960 to 2004 one of "anxiety and doubt," during which the French "have had to come to terms with the legacy of the Occupation, with the loss of empire, with the influx of foreign immigrants, with the rise of Islam, with the destruction of traditional rural life, with the threat of Anglo-American culture to French language and civilization" (1). In that context, Greene emphasizes the importance of 1960s Paris, with its continuous movement, traffic, and speed, which could be captured by the new cameras and editing techniques (248).

Whereas Michel Marie includes both Agnès Varda and Jean Rouch in his account of the nouvelle vague (50, 80, and 96), Greene counts them along with Chris Marker as makers of the contrasting, explicitly political films – "so-called Left Bank and *cinéma vérité* film-makers" (249). The differentiation is important because it allows us to distinguish between the more political films associated with Paris and films that negotiated cultural–political changes through the seemingly apolitical discourse of love. We will see that Agnès Varda's *Cléo from 5 to 7* (1962) articulates politics through a love story mapped onto the cityscape.

The defining moment of the New Wave

The nouvelle vague was a reaction to the tradition of quality ("la tradition de la qualité"), which was characterized by high production value, primarily literary adaptations, high attendance and earnings at the box-office, and a limited number of filmmakers making most of the films. The French New Wave's reaction to this studio system in France was part of a general historical shift from studio production to *auteurism* in the 1960s. The nouvelle vague officially lasted for only two seasons, 1959–60, but it had a lasting effect on later French and international films in that *auteur*-centered cinema also developed in the United States, Germany, Great Britain, Brazil, Japan, Poland, and the Czech Republic. But only in France was the nouvelle vague so closely tied to one city: Paris.

So the term 'nouvelle vague' refers to a time frame that was surprisingly short considering the movement's impact. Marie attributes three primary characteristics to it: it was a coherent movement, it lasted a limited period of time, and simultaneous interlocking factors at the end of the 1950s led to its emergence (2). The term was originally coined not in relationship to film, but as a "sociological investigation of the phenomenon of the new postwar generation" (Marie 5), and then the French newspaper *L'Express* applied it to the new films that defined themselves in contrast to the "tradition of quality," the dominant cinema of the 1950s: 1958 brought Claude Chabrol's film *Handsome Serge*, followed by his *The Cousins* (1959), and in May 1959 François Truffaut's *The 400 Blows* arrived in the theaters and was shown at the Cannes Film Festival. In March 1960 Jean-Luc Godard's film *Breathless* was released. All four films were highly successful with the audience and basically mark the beginning of the nouvelle vague – and all four are set in Paris.

The nouvelle vague relied on a close relationship between criticism and filmmaking – that is, the films were anticipated and accompanied by manifestos by film critics who often became directors themselves, which circumscribed the movement's parameters. *Cahiers du cinéma*, the film journal with which these critics-cum-directors were associated, was first published in April 1951. The so-called "cult of the director" refers not only to the participants in the nouvelle vague, but also to the directors that the *Cahiers du cinema* retroactively identified as *auteurs* whom they tried to emulate: "Jean Renoir, Robert Bresson, Jean Cocteau, Jacques Becker, Abel Gance, Max Ophuls, Jacques Tati, Roger Leenhardt" (Wilson 14). Several of the key concepts of the nouvelle vague were articulated in published articles even earlier than that. For example, according to Marie, Alexandre Astruc's essay "The Birth of a New Avant-Garde: La Caméra-stylo" in *L'Ecran français* in 1948 initiated the French New Wave and offered the "first affirmation of the notion of the *film auteur*" (33). In January 1954 Truffaut published his crucial essay "A Certain Tendency of the French Cinema" ("Une certaine tendance du cinéma français") in *Cahiers du cinema*, arguing against the "tradition de la qualité" (Wilson 13). During the 1960s some films, such as Chabrol's *The Good Girls* (1960), were financial and audience failures, while others, such as Godard's *The Little Soldier* (1960), was banned because of its depiction of the Algerian war. Some more successful films followed, including Godard's *A Woman Is A Woman* (1961) and Truffaut's *Shoot The Piano Player* (1960).

The New Wave valued small budgets, which favored the creative freedom of *auteurs*. Directors tended to rely on "plan-of-action scripts," which are "more open to the uncertainties of production, to chance encounters, and ideas that suddenly come to the *auteur* in the here and now of filming" (Marie 77). This understanding of film-shooting has much in common with Georg Simmel's and Walter Benjamin's understanding of the city (discussed in Chapters 1 and 2), particularly Benjamin's theorization of the *flâneur* as a figure that is seduced by the city to venture into it. He shared with Simmel the emphasis on chance encounters that for Simmel had to be controlled by exact planning so as to prevent the underlying chaos from erupting. In contrast, Benjamin's texts about the *flâneur* situate the pleasure of the city precisely in the chance encounters that are available to the *flâneur*, the figure that leisurely walks through the city. The many tracking shots of the city seen through the eyes of either the characters or the camera capture the sensation of the *flâneur* and the ambivalence of post-Second World War modernity, which was also nostalgic for its own past as embodied in the nineteenth-century Parisian architecture and urban planning that dominates the setting of the nouvelle vague films.

Chance encounters in the city determined the course of shooting and ultimately a film's narrative. The emphasis on the authentic *mise-en-scène* makes the nouvelle vague particularly well-suited to the articulation of urbanity. Marie suggests that because the scripts of the nouvelle vague were more personal, "it was really in the *mise-en-scène*, the relation to the characters, and the serious or ironic private film references that this subjectivity was inscribed" (75). The biographical dimension projects subjectivity onto the urban space created in the films. Paris, however, functions also as the capital of France. The biographical relationship with Paris therefore also results from its role as educational, political, and cultural center of the nation. *Auteurs* imbued the setting with subjectivity, but the relationship between their subjective portraits and the city under investigation was shaped also by the role the capital played for the nation at that historical juncture.

Urban–rural: old binaries in a new dress

Chabrol's film *The Cousins*, considered to be the beginning of the nouvelle vague, relies on character and thematic constellations of urbanity and rurality, sexuality and innocence, decadence and hard work, success and failure, thus continuing the rural/urban split from the nineteenth century. It is the story of 23-year-old Charles, who moves to Paris to live with his cousin Paul and study at the university; and Paul proceeds to introduce him to the world of decadent students. The main setting is Paul's apartment in Neuilly, where he has sex parties. Charles falls in love with Florence, who begins a sexual relationship with Paul. Whereas Charles works hard for his exams and fails, Paul parties and passes. Subsequently, Charles decides to commit suicide by jumping into the Seine, but at the last moment does not to do so. Instead, he gets a gun and tries to kill his cousin while he is sleeping. He pulls the trigger but nothing happens. The next day Paul plays with the gun, and it goes off, killing Charles.

The Cousins continues the city themes established in Weimar city film and film noir. Neuilly-sur-Seine is presented as a wealthy suburb of Paris – it is now associated with conservative politician Nicolas Sarkozy, current president of the French Republic. Paul takes Charles to clubs where young people hang out, smoke, and listen to jazz, and where women proclaim that they are independent of men. The film's first bar scene does not advance the narrative, but instead captures the idle atmosphere among the young people. The seemingly random shots anticipate *cinéma vérité* (cinema of truth), more closely associated with the films of Jean Rouch. The students' decadent lifestyle contrasts with the subplot involving a bookstore owner who advises Charles that in order to succeed he must work hard.

The contrast of city and country organizes the narrative. Charles narrates events in letters to his mother, which the audience "reads" in his voice-overs, so that writing as a traditional form of communication is aligned with the countryside and also with the innocent love of a son for his mother. The city becomes the setting for sexual relationships shaped by power and pain. The intimate perspective of Charles creates a tragic narrative through which the audience is drawn into the city but is simultaneously alienated from it – attracted to and seduced by the whirlwind and sometimes bizarre action of the film, but also sympathetic to the desperate morality of the main character. This underlying narrative structure, despite all its differences, is reminiscent of the relationship of the main male character with the city and its women in film noir.

Greene explains the immense impact of the French New Wave as a generational break: "the young people in New Wave films exhibit a disregard for the social conventions – particularly those governing sexuality – that marked their parents' generation," but as a consequence "the characters in these films are often vulnerable and alone," as is Charles (2004: 248). While walking with Florence and circled by the camera, Charles explains that he has an inferiority complex because he is from a small town in the country. The character of Florence represents the urban love that is detached and immoral, in stark contrast to the invisible mother, the addressee of the letters. The narrative is episodic, similar to the urban experience described by Simmel: scenes show Charles working, Paul playing, and Florence lounging on the balcony. The apartment becomes the set for a drama motivated by the loose morals of the city: the erotic and violent triangle of two men and a femme fatale, here in a contemporary, cool environment.

The neighborhood: affective urbanity

Films such as *The Cousins* rely on a sense of urbanism in which conventional mores and social ties between friends and family have disappeared. However, in the films of the nouvelle vague we also find that the city, or more precisely the neighborhood, appears as the setting for affective relationships substituting for conventional family structures: coffee-houses, bars, and the street become home to young boys, as in Truffaut's *The 400 Blows* (*Les quatre cents coups*, 1959). The city as a whole offers a liberating education to the young main characters of the film, and the film celebrates particularly the neighborhoods Truffaut knew as a child (Greene 2004: 248). *The 400 Blows* depends on the tension that arises between the human drama in the interior, domestic space and the exterior context of the city, which manipulates love.

Truffaut directed a series of films beginning with *The 400 Blows* (1959), *Antoine and Colette* (1962), *Stolen Kisses* (1968), *Bed and Board* (1970), and *Love On The Run* (1978) and extending to a total of four shorts and 21 feature-length films during his lifetime. This chapter focuses on *The 400 Blows*, *Antoine and Colette*, and *Stolen Kisses*, which share an emphasis on love set in Paris, are intertextually related, and belong to Truffaut's earlier oeuvre – clearly part of the nouvelle vague. All cast Jean-Pierre Léaud in the title role as Antoine Doisnel – in *The 400 Blows* as a young boy, in *Antoine and Colette* as a young man, and in *Stolen Kisses* as an older young man.

Much has been made of the biographical dimension of *The 400 Blows*, casting the figure of Antoine Doisnel as the alter-ego of Truffaut, steeped in the urban environment of Paris. The beginning of the film favors the city over the individual in an extended sequence showing Paris; continuous traveling- and tracking-shots center on the Eiffel Tower and move through the area of the Cinémathèque Française. The lightweight camera made these kinds of location shots possible, and they announce and celebrate location shooting in contrast to studio filming. The three-minute opening consists of five shots edited together with the camera tilted slightly up, moving along façades of nineteenth- and early twentieth-century buildings. According to Andre Bazin, too many edits in a scene makes a film seem artificial, and this opening shot includes only a limited number. The Eiffel Tower is centered in the first shot, and the camera is moving towards it. In the next five shots, the Eiffel Tower appears and disappears until the camera moves closer, under the Eiffel Tower, and then picks up speed and moves away, repeating the composition of the very opening shot in reverse, reversing the gaze, and picking up speed in a straight-on shot.

Like an overture, the three-minute segment introduces the theme of the city and evokes a mood with classical, melancholy music on a xylophone and a harp, reminiscent of childhood. No characters are present, no narrative hinted at. The city is empty. Instead of an establishing shot, we see the city from the perspective of a moving car, recreating the experience of the city as modernity not through the eyes of the *flâneur* but from the perspective of the car. Kristin Ross argues that during the 1960s, 1970s, and 1980s France transformed an imperial employment of technology and discipline in the colonies into a concern with technology at home expressed in the love of automobiles, household appliances, and factory production. The opening shot captures celebratory nostalgia that cinematically recuperates the past into the present by capturing bygone youth in the city.

The 400 Blows tells the story of 14-year-old Antoine, who lives with his parents in a small apartment at the Place de Clichy. At school he is mischievous and skips classes to go to the movies or roam through the city with his friend René. At home he has chores to do and hears his mother and stepfather fight. One day he sees his mother with her male lover in the street. When Antoine needs a note from his parents to explain his absences, he forges a written excuse that his mother has died. When his parents find out, they come to school and slap him in front of his classmates, so he does not want to return home and stays with his friend René instead. After he steals a typewriter from his stepfather's workplace, his stepfather hands him over to the police, and Antoine is sent to an institution for delinquent boys until, during a soccer game, he escapes and runs off to the ocean. According to interviews with Truffaut and Léaud, Truffaut did not write a detailed script but instead gave the characters ideas, concepts, and feelings to transmit. In his own words, he created a documentary feeling.

Urban sites, such as streets, movie theaters, and arcades contrast with interior shots of the classroom and the apartment, which are limited and oppressive. At home Antoine has to set the table, stoke the stove, go shopping, and take out the garbage. His mother yells at him, and both parents are suspicious of him and try to control him, scolding him for lying. In contrast, Paris becomes a great playground for the youthful adventurers. The scenes at home and at the school are not accompanied by a soundtrack, but the scenes in the city are accompanied by a playful, upbeat piano score that ensues anytime a new day starts and Antoine and René skip school. The wide-angle shots situate the two teenagers in the streets among other people who pay no particular attention to them. Episodically we see them going to the movies, crossing and walking along the streets, and playing pinball machines in the coffee-houses. One humorous scene shows from a high angle the physical education teacher running with his students through the city. He gradually loses his students in groups of one to three until he is followed by the last three remaining – a wonderful illustration of Benjamin's views on the seduction of the city.

The main characters, all children, inhabit the city with ease; their mischief is fed by the city's urban entertainment. While Antoine is betrayed by the adults around him, the childhood friends take care of each other. Attachment and affection mark the process of remembering childhood in the city, while love between adults is marked by betrayal of an unreliable woman. In a scene at a carnival, Antoine stands in a machine that spins around so fast that it creates an effect like early moving pictures and that also includes a cameo appearance by Truffaut. By putting himself, the director, in what looks like an oversized zoetrope, Truffaut actively positions himself in the history of cinema and simultaneously celebrates its magic. Truffaut's cameo appearance here is also an explicitly self-referential gesture about the relationship between urban spaces – the carnival – and the cinema, and points to remembering and documenting as processes that link the two. The cameo also casts Truffaut's biographical claim as a highly self-aware and self-referentially mediated account: he shows through his body that he is both subject and object of the film, meeting the present embodiment of his past self in the machine of moving images. *The 400 Blows* connects here to the early history of film and cinema's potential for manipulating time and place, as outlined in the Introduction.

Open windows: permeable interior and exterior spaces

Truffaut continued the topic of love in the city in the short film *Antoine and Colette* (1962), part of an international episodic film entitled *Love at Twenty*. Léaud continues to play Antoine and grows up with him, mimicking the conventions of a long-term documentary. Antoine has returned to Paris and lives in a small room in the Hôtel de la Paix (at the Rue Forest), next to the Gaumont Palace movie theater; his window overlooks the Boulevard de Clichy.[8] He works at Phillips, making records. During a classical concert at the Salle Pleyel he falls in love with Colette and subsequently moves into a room in the Hôtel de l'Europe, across the street from her family. He is invited to eat and watch television with her parents while she goes out with other young men.

Like *The 400 Blows*, *Antoine and Colette* opens with Paris itself before we are introduced to the main character. A shot of a busy intersection begins with a clock at the brasserie Le Bastons, and "Paris" is written across the screen, in part presumably because the omnibus film was shot in five different countries. The name François Truffaut appears in the center of the screen, written onto the busy Parisian intersection, and then the camera pans to the left across the Gaumont Palace, which is advertising *The Count of Monte Christo*, to the Hôtel de la Paix and up to Antoine's room. Only then does a cut show a close-up of the alarm clock next to his bed. The *mise-en-scène* of Antoine's room contrasts with the oppressive quality of the interiors in *The 400 Blows* in that it is decorated with record sleeves.

Then he opens the shutters onto the Boulevard de Clichy, linking his awakening to the view of the city and to his sexual awakening, announced by the title of the omnibus film.

The interior life of Antoine is integrally intertwined with the exterior life of the city via the *mise-en-scène*. He opens the windows onto the street after putting on a classical music record, which provides the soundtrack for a panning shot that shows us the Boulevard de Clichy over his shoulder. Then a shot/reverse shot shows him standing on the balcony, relaxed and smoking as the camera zooms in, and when the camera returns to the street, the voice-over tells his story, the repeated shot/reverse shots having established his intimate relationship with Paris. When Antoine runs to catch the bus he takes to work, the film echoes the children's movements in *The 400 Blows* as movement motivated now by a mature lifestyle. In *The 400 Blows* Antoine and René walked, ran, and played in the city, and now Antoine runs to work and clocks in. After work, he meets René and they go to a classical music concert at the Salle Pleyel, a mature variation on the experience of the cinema in *The 400 Blows*. Antoine is characterized as a true urbanite, embedded in the urban space, the capital of France, productive in the sphere of culture, but without the social network of the traditional family.

Antoine notices Colette at the concert, and he and René follow her and her girlfriend down the stairs to the subway. The encounter is repeated during the week, and Antoine concludes that she must live in his neighborhood. They meet and exchange addresses, and continue to meet at coffee-houses and at lectures in the city. A voice-over narrates their encounters and claims that Colette treats Antoine like a friend, but that he does not notice or does not care, as if he has no interiority and all his desires are expressed through the exteriority of the city and its possibilities. Camera and narrative position the spectator to identify with Antoine, and as he follows Colette repeatedly, the *flâneur* turns into the voyeur and the woman becomes the object of desire to be "hunted" in the city – as René expresses it when Antoine tells him about the three encounters with Colette: "Preliminaries are over. Time to attack."[9] However, Colette's sexuality is corrupted by the city and she is not to be caught. During a phone conversation Antoine asks why she did not come to a concert, and she explains that she ran into friends and that they partied all night; she avoids commitment instead of answering his love letters.

One day, when Antoine attempts to visit Colette, who is not home, we see him in the hallway, in an over-the-shoulder shot that emphasizes a window open into the street, the kind of shot used repeatedly to emphasize the on-location shooting in the city and also to create a fluid sense of urban space between exteriority and interiority. Unlike in *The 400 Blows*, interior shots do not capture a space absolutely separate from the exterior, and the lives of the characters participate in the fluidity between inside and outside. This fluidity is expressed also in Antoine's movements, as when he leaves his apartment, runs right out into the streets, sees another hotel across the street, the Hôtel de la Paix, and moves in. The idea that he is part of the city is reinforced by his move from one hotel to another. In the Hôtel de L'Europe he is standing at the window and watching the street when Colette's family arrives. The shot repeats the film's opening shot/counter-shot, zooming in on Antoine and thus signaling the relationship between him and the city, though the familiar faces of Colette's family have replaced the urban anonymity.

Antoine and Colette concludes with a sequence of photos by Henri Cartier-Bresson and the title song *L'amour à vingt ans*. The black-and-white photos show various couples kissing and in love in Paris, shifting from the individual narrative of Antoine to the paradigmatic representation of Paris expressed by Cartier-Bresson as a city of young love. Truffaut's portrayal, however, shows us a frustrated love fed nostalgically by affection for Paris.

The street: love and detection

The New Wave, observed Éric Rohmer, "was born from the desire to show Paris, to go down into the street, at a time when French cinema was a cinema of studios" (1981: 34; translation Greene 2004: 248). The next installment of Antoine's story, *Stolen Kisses* (1968) begins, like *The 400 Blows*, with the conjoining of a personal inscription by Truffaut and the city of Paris. The very first shot shows us an unknown street with the first credit item, the production company, written across it: "Une Production/Les Films Du Carrosse/Les Productions Artistes Associés." Then the title of the film appears, followed by "Dedicated to Henri Langlois, the Head of the Cinémathèque Française," and a clearly centered shot onto the Cinémathèque Française with a handwritten sign announcing that the date of reopening will be announced in the press. At that point Henri Langlois, the Head of the Cinémathèque Française, who had built an unrivalled film collection by hiding films from the Nazis during the German occupation and had created an intellectual and artistic center for the nouvelle vague in the Cinémathèque, had been fired, which led to an outraged protest by the filmmakers and intellectuals associated with it. The camera zooms back and, after the cut, a traditional establishing shot of Paris shows the Eiffel Tower dominating the skyline. Truffaut's name is written across this image, again aligning his name with the dominant signifier of the city. While the camera pans to the left and down, the nostalgic *chanson* about remembering love from the credit sequence ends, and we hear the real sounds of the city streets. The camera moves towards a small, barred window, and we hear a neutral dialogue before a second cut deposits us in the interior setting, a military jail.

Again the narrative is set in the city of Paris and carries clear biographical elements, which tie Truffaut's life to the city. But the love stories that structure *Stolen Kisses* are shadowed by another aspect of urban narratives, namely the convention of the detective story. Among his other jobs, Antoine works as a private detective, adding another layer to the constant movement throughout the city that characterizes Truffaut's films about Antoine. The trope of detection, with its dimension of voyeurism, refracts the understanding of love associated with Paris. While *The 400 Blows* portrays innocent play in the city and *Antoine and Colette* confronts the sexual awakening of a young man with the liberated woman, *Stolen Kisses* offers a rather comedic staging of the professionalization of moving and observing in the city. This process of maturation in relation to the city across several films captures the urban dweller as a masculine subjectivity, which culminates in the role of the urban detective.

Reversing the gaze: the female *flâneur*

The observation just made about *Stolen Kisses* is echoed by Emma Wilson's general comment that "the Parisian landscape of the *nouvelle vague* is a space largely of male subjectivity and of amorous encounters" (124). In Chabrol's and Truffaut's films, the city is gendered as male: men inhabit the city and actively move through it, while women serve as objects of their desire. At the same time, the status of women was changing and female characters were portrayed as sexually liberated and thus often as sexualized, sometimes as dominant characters in contrast to earlier representations in French cinema that emphasized traditional feminine qualities.

We find an important counterpoint to the films discussed so far in Agnès Varda's *Cléo from 5 to 7* (1962), in which the main character, *chanson* singer Cléo, awaits the results of a hospital medical exam. In almost real-time fashion, the 90-minute film shows us Cléo at a tarot reading, then with her friend and maid, Angèle, at a coffee-house, at home with her male visitors and composers, in the streets of Paris with her friend Dorothée, and then alone in the Parc de Montsouris, where she meets a soldier of the Algerian war who is on leave and accompanies her back to the hospital to get her results.

Janice Mouton rightly points out: "Among the pleasures of viewing *Cléo from 5 to 7* (1961) are the scenic views of Paris" (3). Repeated and very long takes show Cléo walking through the city streets, alone or with her friends, watching little scandals and activities that attract crowds. Extended scenes show her driving through the city, also alone or with friends, in cabs, cars, and finally on a bus. Urban activities include running errands, buying a hat, and watching a short, black-and-white, silent film in a movie theater, which has cameos by Jean-Luc Godard,

Anna Karina, and Jean-Claude Brialy. Janice Mouton has argued that Cléo's transformation into a female *flâneur* is "rooted in her direct involvement with the city," that Paris "both responds to and structures her flâneuristic activity" (14).

In contrast to windows as an organizing motif in Truffaut's Leaud films, the *mise-en-scène* in *Cléo* is structured by mirrors, a feature that reflects the gender of the main character in a gendered view of the city. After leaving the tarot-card reader and arriving in the front hallway of the house, Cléo looks into a mirror and we hear her voice-over: "As long as I am beautiful, I am alive." Mirrors are used repeatedly to create split screens. In the first scene in the coffee-house, Angèle and Cléo sit in front of one that splits the background, one half reflecting the young women, the other half creating the depth of the coffee-house. Cléo turns around to talk to herself in the one half; the other extends the interior space, so that the waiter and the owner and parts of the street are reflected. Mirrors on the back wall expand the space and multiply the characters. Through the split screen we see strangers on the right having a fight which ends with the man leaving – effectively creating a space in the city in which private stories are on public display. The mirrors and the glass of the walls have the effect also of creating the fluidity of interior and exterior that characterized Truffaut's films. For example, when Cléo is attracted to a hat in a milliner's store, the first shot is from inside the store, creating fluidity between

Figure 3.1 Agnès Varda. *Cléo from 5 to 7* (1962): Production still with Agnès Varda in the mirror

there and the outside. Inside the shop, several mirrors allow Cléo to look at herself, and her spontaneous purchase of a hat characterizes the city as a space of seductive commodities available particularly to women.

The film shows Cléo's development from "feminine" consumption of commodities (the stories of Antoine, René, Charles, and Paul include episodes of consumption only of women and the city) to a genuine human encounter at the film's conclusion. But the city is also a public space in which news, particularly of war and student activity, circulates. During the long cab-ride back to the hospital (with a female cabdriver), Cléo is confronted with African masks, art students surrounding the car in costume, and radio news of the Algerian war, elements that not only serve as a backdrop in the film, but continue to haunt Cléo's memories. The camera shifts between viewing Cléo and point-of-view shots from her perspective, emphasizing the interaction between the city and the female gaze, even though for long stretches any subjective point of view is withheld.

The very first view of Cléo moving through space, when she leaves the apartment of the tarot-card reader, sets up the convention of the over-the-shoulder shot of her

Figure 3.2 ***Cléo from 5 to 7*****: Art students approaching the cab**

going down the stairs, then a medium close-up of her face as she is moving, and finally a point-of-view shot once she is looking out the window into a courtyard. These three steps align the spectator with Cléo's perspective on the city, reinforcing her gaze. The long takes of her traveling from the house of the tarot-card reader to the café where she meets Angèle uses only one, almost invisible, cut, creating a fluid movement of Cléo through the city.

Mouton suggests that Cléo undergoes a transformation from "feminine masquerade to flâneuse," enabled by the city of Paris itself (3). The many shots of her looking into mirrors at the beginning of the film shows that she is so concerned about her own looks that she does not register her surroundings. Mouton interprets Cléo's statement "As long as I'm beautiful, I'm alive" as indicative of her "feminine masquerade," a term that was initiated by Joan Riviere and taken up by Mary Ann Doane. In Riviere's account, "feminine masquerade" refers to successful women who act in a stereotypically feminine manner to avert the threat of castration (in psychoanalytic terms) that they represent to men. Mouton suggests that Cléo's shopping for the hat is not just a story of "commodity fetishism in the Marxist sense," but instead offers a moment where "fetishism and feminine masquerade converge," because Cléo desires these objects to "adorn her body, transforming her into a fetishized object" (6).

The film's turning point occurs when Cléo has sung her *chanson*, *Cri d'Amour*. The screen goes black but is revealed to be a curtain, which Cléo draws aside and steps through, dressed in black; she takes off her blonde wig and leaves the house for the streets of Paris. As on her departure from the tarot reader, she finds a mirror in the street but observes differently: "My unchanging doll's face./ This ridiculous hat./ I can't see my own fears./ I thought everyone looked at me. I only look at myself./ It wears me out." In this moment of critical recognition of her subjectivity in relation to those around her, Mouton locates her transformation into a "flâneuse" (7).

Paris is central to Mouton's argument, because it is the interaction with the city that turns Cléo, first, into an "observer of the crowd," and then into "part of it" (9). Mouton suggests that Varda chose the locations of the city deliberately to portray the unfolding of Cléo's new subjectivity and because they invite female flânerie: "the rue de Rivoli, the busy shopping street; the rue Huyghens in Cléo's neighborhood; the café Le Dôme, on the Boulevard Montparnasse, where all of Paris passes by; and the Parc Montsouris, with its mix of people and refreshing natural space" (9). And a series of episodes shows Cléo's development: in the Montparnasse café Le Dôme, Cléo notices details about others, while the short silent film Cléo watches with her friend Dorothée is also "about learning to look, and about the potentially transformative power of looking," because the character in the short film, on a city street, wears dark sunglasses and misreads the situation (13).

The last episode, importantly, is the encounter with the soldier Antoine. *Cléo from 5 to 7* does not portray a traditional love, but rather a chance encounter between a *chanson* singer – the embodiment of French culture – and a man who is leaving that very day to return to the Algerian war. This random encounter in the city is the memorable and affective kernel of the narrative, in contrast to Cléo's meeting with her lover, José, who discovered her talent but now only stops by her apartment. Meeting Antoine in a public park validates the potential emotional depth of random encounters in the city. Whereas in Benjamin's theory of the *flâneur* and many of the Weimar city films seductive commodity and object of desire are collapsed in the figure of the prostitute, Varda's film pries that conflation apart and allows us the pleasure of the chance encounter in the city.

Case Study 3 **Jean-Luc Godard's *Breathless* (1960)**

Jean-Luc Godard, one of the most influential filmmakers of the nouvelle vague, was born in 1930 in Paris. He participated in the *ciné-club* movement, which was a network for the distribution, reception, discussion, and theorization of cinema that included André Bazin, Jacques Rivette, Claude Chabrol, François Truffaut, and Jacques Demy. When André Bazin founded *Cahiers du Cinéma* in 1951, Godard was among the first writers for the important new film journal. Before he made his first feature film in 1960, *Breathless*, which was to become the seminal work of the nouvelle vague, Godard shot several short films.

Breathless depicts a love affair between Michel Poiccard, a French small-time gangster, and Patricia Franchini, an American student in Paris who works for the American newspaper the *New York Herald Tribune*. The narrative turns the cityscape into a paranoid space echoing the conventions of film noir, and Patricia's betrayal of Michel reflects the femme fatale's betrayal of the main male character in film noir. Wilson emphasizes that Godard shot his film in real locations to "establish his filmic city as a real geographical space: the Champs Elysées where Patricia sells her *New York Herald Tribune*, Notre Dame viewed from a passing car, the Arc de Triomphe and the Tour Eiffel" (72). The technique is a primary characteristic of the nouvelle vague and was made possible by the development of the light, moveable camera, here wielded by Raoul Coutard, described by Wilson as "one of the most celebrated cinematographers of his generation," and whose flexible

Figure 3.3 Jean-Luc Godard. *Breathless* (1960): Michel and Patricia on the Champs Elisées

camera technique enabled the film to capture the playful, improvised movement of the two main characters (72).

Wilson calls *Breathless* "the film which best articulates France's relationship to film noir [as] transatlantic exchange" (70). Like Lang's *M*, *Breathless* maps its manhunt onto the streets of the city and positions the film in a line from Weimar to film noir, which allows it to revise the notion of romantic, cinematic love. Paris is inhabited by Americans and American culture; in Greene's opinion, "Paris became less French," paradoxically precisely at the moment when it was cinematically represented as real and authentic (2004: 248).

Breathless begins, however, not in Paris but in Marseilles, where Michel steals a car and leaves behind his female accomplice. We see him driving along a chaussée in a point-of-view shot accompanied by music reminiscent of film noir. The opening scene harks back to B-movies, characterized by action that is exaggerated and formulaic, referencing self-understood narrative conventions. When Michel is chased by the police, he follows a little road into the woods and shoots

continued

the police officer with a gun from the car's glove box. The sequence's editing does not follow the rules of continuity editing. We see the cop approaching on his bike, Michel bending over into the car, we hear him say "Move or I shoot you" over a close-up of the gun and we see the police officer fall to the ground in the woods, and then a long-take shows Michel running away. The film's aesthetics cite film noir of the B-category in stark contrast to the studio-based, high-production value, star-driven, literary adaptation of French cinema that came before the New Wave. Whereas French, national, high culture traditionally viewed the country's landscape as pastoral, the narrative and aesthetics in *Breathless* reinterpret the French landscape as an alienated, hybrid setting.

The film explicitly foregrounds the relationship between landscape and subjectivity in the improvised and disaffected monologue during which the actor, Belmondo, addresses the camera in an ambiguous way. The shot leaves open whether the fourth wall is broken and he is intentionally addressing the audience, or whether Michel, the character, is talking to himself. He says to the camera: "Nice countryside./ I'm very fond of France./ If you don't like the sea,/ if you don't like the mountains,/ if you don't like the city,/ then get stuffed!" While classical cinema is based on the attempt to cover its tracks through continuity editing, *Breathless* emphasizes the presence of the camera, as when it turns to two hitchhikers on the street and then returns to Michel. The camera becomes independent, dislodged from character and narrative, neither an illusion of an omnipotent narrator nor Michel's subjective point of view.

The improvisational script is translated in recognizable ways into Michel's character: he talks to himself, sings without words, makes up words and songs, in a way that neither constitutes significant dialogue nor is enunciated clearly for the camera. The nouvelle vague preferred as actors amateurs or unknown professionals. According to Ginette Vincendeau, *Breathless* made Belmondo famous, paradoxically, because he was almost entirely unknown at the time of the making of the film, which granted authenticity (165). Vincendeau emphasizes his masculine "cool," which, she argues, was instantiated by his performance in *Breathless*: "Emerging from behind a newspaper in the first shot, Belmondo, with his dangling cigarette and casually insolent delivery, was to symbolize 1960s cool ever after" (161).

Vincendeau interprets a "key scene" in *Breathless* when Belmondo looks at a photo of Humphrey Bogart's face at a cinema on the Champs Elysées and "models his expression on Bogart's: slight frown, 'sad' eyes, introspective stance, a way of holding his cigarette and of rubbing his upper lip" (164). The "sense of authenticity" is also associated with Belmondo's face, which is "irregular, elongated, with thick lips and nose flattened by boxing, and, from early on, expression lines" (166). His "cool" and aloof performance also ushered in a new notion of sex appeal that integrated the urban detachment of the gangster or private eye from the film noir tradition. In contrast to Michel, Jean Seberg's Patricia is "audibly not French. As a result, Belmodo's Frenchness comes into relief" (Vincedeau 166). In addition, the film shows his nude torso and offers him as a masculine object of desire in a long scene on a bed in a hotel room. His body and his cool are intimately tied to the city of Paris.

By appropriating representative images of American cool for a European urban artistic subjectivity marked by the *mise-en-scène* of Paris, the film reinterprets the national narrative of Frenchness and the jazz score aurally reinterprets Paris. Michel enters and leaves coffee-houses, runs across streets, scans newspapers, visits women, and scans their purses for money. He has been in Rome, he asks his friend whether Gabi has returned from Spain, he is in love with a woman from the United States – characters are rootless. Transitional places, like hotel rooms rather than family homes, provide the *mise-en-scène* for interior spaces. The characters live transitional lives in contrast to Paris's monumental architecture.

Both Michel and Patricia are displaced and uprooted characters. *Breathless* shows important sites without mobilizing traditional cultural significance, and thus reinterprets Paris as open space and an open signifier. An extended hotel-room scene is at the heart of the film. Improvisation results from limited space: the characters change clothes, kiss, cross over and around the bed, talk on the phone, talk with each other, and look at pictures. The interior spaces in Paris are transitional, without roots, culture, family, or context. Michel enters Patricia's hotel and her hotel room without being registered. She sits on the bed with sunglasses, he with a hat, cigarette, and sunglasses. Their fleeting kisses reinterpret the love story convention. Michel's real name is Leszlo Kovacs, revealing him as having Eastern European connections. When he is in a cab with Patricia, he lies about an accident and points to a house where he says he was born. The film leaves open whether he has created an imaginary French identity for himself, anchored in Paris, or whether he actually is Parisian.

Patricia is a different kind of streetwalker from the figure of the prostitute who populates Weimar cinema. As in *M*, the city itself announces the imminent arrest of Michel Poiccard, displaying the narrative as urban text. The two characters end up in yet another transitional apartment with a map of Paris on the wall; Michel gets killed, a close-up of Patricia shows her reaction, and then she turns away.

continued

Breathless is a love story and an anti-love story, alluding to all the generic conventions and undermining them. This new kind of cool love story is intimately tied to the city space of Paris and thus rewrites Paris.

Further reading

Dudley Andrew (ed.) (1987) *Breathless: Jean-Luc Godard, Director*, New Brunswick, NJ: Rutgers University Press. The books in this series include the film's script; an Introduction by the book's editor; and other documents, such as the original treatment, interviews, reviews, a filmography, and a bibliography. They provide a invaluable source of information for discussing individual films.

Sandy Flitterman-Lewis (1996) *To Desire Differently: Feminism and the French Cinema*, New York: Columbia University Press. Flitterman-Lewis's book contextualizes selected examples of Agnès Varda's films in the history of women's films in France and in the context of feminist film theory.

Michel Marie (1997, 2003) *The French New Wave: An Artistic School*, trans. Richard Neupert, Malden, MA: Blackwell Publishing. This book gives an overview of the movement in a historical outline, paying particular attention to the manifestos that accompanied the movement and its influence on international cinema.

Richard Neupert (2002) *A History of the French New Wave*, Madison: University of Wisconsin Press. Neupert's book provides a survey organized according to directors to address the movement comprehensively.

Essential viewing

Claude Chabrol. *The Cousins* (1959)

Jean Douchet, Jean Rouch, Jean-Daniel Pollet, Eric Rohmer, Jean-Luc Godard, and Claude Chabrol. *Six in Paris* (1965)

Jean-Luc Godard. *Breathless* (1960)

François Truffaut. *The 400 Blows* (1959)

—— *Antoine and Colette* (1962)

—— *Stolen Kisses* (1968)

Agnès Varda. *Cléo from 5 to 7* (1962)

SECTION II

Section I focused on three important moments in the development of national cinemas to show continuities and breaks in the cinematic portrait of the city. Section II shows the progression from modernist national cinema to transnational cinema, which relies on postmodern narratives and aesthetics. The three chapters are organized according to themes: Chapter 4 discusses the Hong Kong film industry, Chapter 5 the cinematic representation of war-torn and divided cities, and Chapter 6 utopian and dystopian visions of cinematic urbanism. These themes highlight the political and historical dimensions of the cinematic representation of cities. In each of the three chapters the development of film moves from modernism to postmodernism and from national to transnational, the common focal point being the city.

This section shows the different political uses to which portraits of urbanism are being put. For example, in the ideological conflict between communist China and capitalist Hong Kong, the urban is associated with the moral corruption of capitalism and rural life with wholesome tradition. Similarly, most films that focus on war and destruction are political in nature, sometimes advocating resistance or revolution, but most often portraying the devastating results of war on cities and their inhabitants. Even science fiction film, which as a genre appears to be detached from contemporary politics, reflects underlying political attitudes in its utopian and dystopian visions of technology and urbanity.

History is at stake in all three chapters, albeit in different ways. The meaning of the past is thematized explicitly in national cinema engaging with its national history. For example, early Hong Kong action film mythologizes Chinese history in martial arts epics. Historical changes can also enable a turning-away from history as in postmodern, transnational cinema, for example, when Hong Kong-born urbanites, detached from mainland China, have integrated martial arts with urban genres such as the gangster film in a unique hybrid genre, the so-called "heroes cycle." National and international history also deeply shape films about wars, civil wars, revolution and their effects on urban terrains. Certain genres, such as the "rubble films" shot in the ruins of immediately postwar Germany, rely on the destroyed cityscape to claim their particular historical moment. In contrast, science fiction as a genre advances its own fantasy of independence from specific historical moments, but as Chapter 6 shows, utopian and dystopian visions of urbanity are intimately tied to historical developments at the time of the films' production.

Each of these three chapters also engage with different configurations of national and transnational production and circulation of films. Chapter 4 accounts for the interplay of national cinema and its transnational context. Hong Kong's most famous martial arts star, Bruce Lee, answered the nostalgic longing of a diasporic Chinese community for a lost homeland, while his heroic film persona invited transnational identification. National and transnational identification are not always at odds with each other, however, not least because their simultaneous appeal is of financial interest to production companies. Similarly, films on war sometimes appear to advance a national discourse but are based on transnational production, as in René Clément's *Is Paris Burning?* (1966). More often they address a national audience, even to the degree of advancing a nationalist agenda. The three chapters delineate subtle historical shifts from national cinemas to transnational cinematic practices.

The underlying argument here is that the national and transnational cinemas parallel modernist and postmodernist aesthetics. Whereas the examples of national cinema are deeply rooted in their individual histories, transnational, postmodern films employ historical references without engaging with their political implications. In the Hong Kong film industry, that shift is exemplified in the two different star personas of Bruce Lee and Jackie Chan. Both have transnational appeal, but Bruce Lee's star persona invokes a national diasporic discourse, whereas Jackie Chan's transnational star persona disavows his specific Hong Kong identity in favor of an undifferentiated general Asianness. Chapter 5 diagnoses a shift from deeply historically embedded films about specific war and postwar moments in specific national settings to postmodern films that rely on loose historical references in post-apocalyptic settings without engaging with the political and moral questions associated with particular historical moments. Chapter 6 tracks the development from early science fiction film, which projects modernity onto the futuristic cityscape, to later science fiction film, which portrays the built, urban environment as dilapidated and out-dated in a postmodern phantasmagoria that associates futurism with virtual reality.

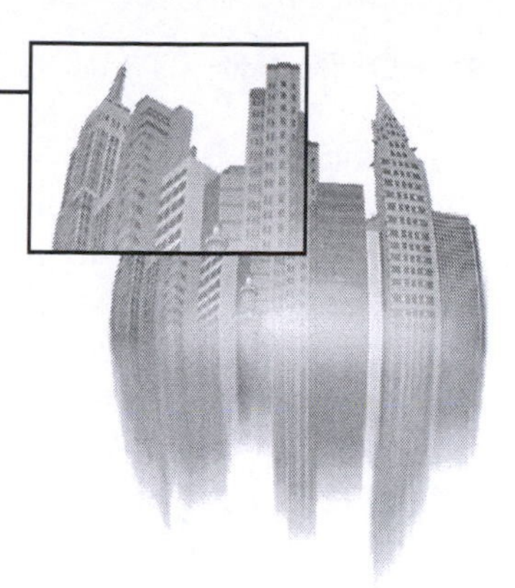

4 City film industry: Hong Kong

> It is probable that this has everything to do with my transplant from Shanghai to Hong Kong at the age of 5. When I got there, I spoke nothing but Shanghainese, whereas Cantonese was, and still is, the local dialect. For some time, I was totally alienated, and it was like the biggest nightmare of my life. It might not be conscious, but certainly I have an intense feeling for geographical upheavals.
>
> Wong Kar-wei

Learning objectives

- To understand the history of Hong Kong in relationship to China and Great Britain
- To situate the development of the Hong Kong film industry in that history
- To contextualize the two versions of martial arts films embodied by the two stars Bruce Lee and Jackie Chan and the two waves of urban films, the heroes cycle and the New Wave, in the particular history of Hong Kong
- To analyze the cinematic representations of Hong Kong in their particular and appropriate urban, regional, diasporic, transnational, and postmodern contexts

Introduction

This chapter continues the discussion of the triangulated relationship of a real particular city, the city–state of Hong Kong, the formation of its film industry, and its imaginary construction by its cinema. Hong Kong cinema is the third largest film producer in the world and frequently outsells Hollywood. Its immense cultural output shows the traces of its unique history in an explicit dialogue with its colonial heritage, its Chinese roots, and its transnational context. Hong Kong culture is shaped by British colonialism, the legacy of China, the simultaneous intentional articulation of separation from China by a diasporic community and, finally, the transnational education, orientation, and business connection of the post-Second World War generation. Hence Hong Kong's film culture is discussed here in relation to mainland China, the region, and transnational global culture. An active, contemporary exchange with Hollywood of actors, directors, martial arts coaches, and choreographers, and the cult circulation among African-Americans and Asian-Americans especially in the 1970s, have characterized the reception of Hong Kong action cinema in the USA.

With regard to organization, while Chapters 1, 2, and 3 discuss films associated with a particular historical moment or period, this chapter casts a wider net and discusses films made in Hong Kong from the 1920s to the late 1990s, beginning with an overview of the emergence of the Hong Kong film industry out of socio-economic conditions created by its particular history. The chapter considers the role that the urban–rural binary plays in the ideological conflict between communist China and capitalist Hong Kong. It then addresses the shift from the martial arts – kung fu – cinema embodied by Bruce Lee to the martial arts comedies associated with Jackie Chan, a shift which coincides with a change from addressing the Chinese diaspora to a transnational audience, and from nostalgia to urban transnationalism. The discussion concludes with the two groups of urban films that have emerged out of Hong Kong and that still continue, the heroes cycle and the New Wave.

A short history of the Hong Kong film industry

Hong Kong was a crown colony of the United Kingdom until the transfer of sovereignty to the People's Republic of China in 1997, and its experience of colonialism and occupation also included the Japanese occupation during the Second World War. Hong Kong's population was very much shaped by immigration from China, which in turn formed a basis for Hong Kong culture. Throughout the twentieth century, the importance of urbanism in Hong Kong film increased in proportion to the length of time diasporic producers and consumers of Hong Kong film had been separated from China. A first shift occurred in Hong Kong action cinema from mythical martial arts narratives tied to Chinese history to urban dramas that subordinated martial arts to a cool, stylized look with fetishized gun violence taking place in Hong Kong itself. A second shift occurred with the Hong Kong New Wave, a trend advanced by young filmmakers educated abroad who create melodramatic art films situated in Hong Kong that stage isolation and displacement in its urban environment. This chapter's emphasis on Hong Kong's film industry presents a methodological shift from the preceding three chapters, discussing its history and the stars, directors, and producers that created the industry and that were created by it.

In *The Asian Film Industry*, John A. Lent outlines the development of the film industry in Hong Kong to become one of the most powerful and active in the world: witness "Asia's largest studio complex," called Movie Town, the high regular attendance at film screenings, and the Hong Kong International Film Festival (92). Li Cheuk-to maps out three phases of modern Hong Kong cinema: the "classical period," 1946–70, which relied on studio production; the "transitional period," 1971–78, which saw the emergence of kung fu and the disappearance and then return of the Cantonese language in the martial arts films; and finally the "modern period," which started in 1979 and led to the New Wave, with a "self-consciously upscale Cinema City look" (quoted in Bordwell 72). David Bordwell suggests that "the production boom of the late 1980s launched a fourth phase. For one thing, it attracted Triads, secret societies originating in China, who now saw film production as not only a money-laundering device but also a reliable source of income" (72).

Beginning in 1923 with Li Min Wei's film *Rouge* and continuing through the 1930s, Hong Kong was the center for Cantonese films, including approximately 100 produced by about 50 companies between 1932 and 1936 (Lent 93; also Jarvie 1977: 9). Ultimately, the expansion of the Hong Kong film industry was a result of political conditions in mainland China, which created two groups of refugees who left, however, for opposite reasons. One group consisted of politically oriented individuals who made films in the official Chinese language, Mandarin, to express opposition to the Japanese. When the war with Japan broke out in 1937, they fled Shanghai. The other group left for Hong Kong precisely because the Kuomintang (KMT, China's Nationalist Party) decreed that films in mainland China had to be made in Mandarin. As producers of films in Cantonese in Hong Kong, they benefited from financial and trading advantages, including the free import of raw film, cheap land for building studios, and limited taxes, regulations, and licenses (Lent 93; also Armes 1987: 158). Subsequently, under the Japanese occupation of Hong Kong during the war, the number of Cantonese films made there increased considerably, and while the film industry continued after the war, production no longer increased, because the political situation in the region remained unclear. The political tension of left and right – the battle between communists and Kuomintang in mainland China – also existed among the film workers in Hong Kong: leftist–communist production companies included, for example, Fiftieth Year Film Company and Feng Huang (Phoenix) Motion Picture Company, and anti-communist companies including the Shaw Brothers and Cathay (Lent 97). Still other film production companies included capitalists *and* communists and produced propaganda *and* entertainment films.

During the 1950s the industry prospered, which Lent interprets as a result of several factors: refugees with "democratic ideas" entered Hong Kong, and new technical developments, such as "deep focus, asymmetrical framing, medium-long shots, and full-stage shots," enabled more cinematically sophisticated productions (Lent 95; also Lin 1979: 15). Steve Fore situates the take-off of the film industry in the 1950s in the context of the larger industrial development of Hong Kong, which "was stimulated by a combination of the post-1949 influx of refugees from the mainland (many of whom arrived in the territory with useful skills and entrepreneurial experience) and the related embargo on Chinese trade by most of the capitalist West" (122). The embargo by the West destroyed Hong Kong's entrepôt status, which had made it an extremely important import–export trading center where no import duties were charged. The end of its entrepôt status, however, created a need for new industries, and additional film production companies were founded, producing mostly swordplays and melodramas.

Joseph Sunn, Lee Tsu Yung, Chang Shin-Kuam, Loke Wan Tho, and Raymond Chow were important in building and maintaining the Hong Kong film industry. The company of Run Run Shaw dominated until 1986, when it stopped production. Run Run Shaw and his three brothers came from a wealthy Shanghai textile family who became involved in film production in the 1920s. In the late 1930s, their empire included theaters and amusement parks in Malaysia, Singapore, Borneo, Java, and Thailand, and in 1958 they relocated to Hong Kong, where in 1962 Run Run Shaw built Movie Town (Lent 98; also Sun 1982: 41):

> [A] 46-acre spread that enclosed ten studios, 16 permanent outdoor sets, three dubbing studios, many film-processing labs, and dormitory and apartment space for staff. The self-contained unit kept 1,500 actors/actresses under contract, as well as 2,000 other staff; maintained its own drama school of 120 students; published periodicals (e.g., *Hong Kong Movie News*) that boosted Shaw stars; and used a wardrobe of 80,000 costumes of all dynasties.
> (Lent 98–9; also Dadameah 1972: 13)

The Shaw studios dominated the Hong Kong market throughout the 1970s, but in 1986 they ended film production because more money could be made in television. Meanwhile, in 1950 Cinema City had been founded by three film-makers, Carl Mak, Dean Shek, and Raymond Wong, who were interested in film as a medium and were backed by strong investments. Cinema City invested particularly in big-budget films such as its big success *Aces Go Places* (1982), directed by Eric Tsang, which paralleled a shift in audience taste to more mainstream and mixed-genre narratives. It also produced John Woo's blockbuster *A Better Tomorrow* (1986), the first of the urban heroes cycle.

Yet another important company, Golden Harvest, was founded in the 1970s by Raymond Chow, who had worked for the Shaw company since 1958 as its publicity manager and head of production. Golden Harvest took on Bruce Lee and began co-producing with Hollywood companies, leading to its kung-fu action films, including the Bruce Lee vehicle *Enter the Dragon* (dir. Robert Clouse, 1973), which "grossed US $100 million in the United States alone" (Lent 100; also Sun 1982: 40). In 1980, Golden Harvest took on Jackie Chan as a star and shifted production by prioritizing the international over the Hong Kong market. During that time, a few other production companies sprang up, several of them backed by considerable capital and interested in film primarily as an investment, including D&B Films, established by Dickson Pon, and Far East Motion Picture Development Ltd., founded by Deacon Chiu.

Until the New Wave, which began in the early 1980s, Hong Kong cinema was generally considered *popular* cinema, which meant it de-emphasized narrative. Whereas in the Hollywood studio system a film is developed from a script, in the Hong Kong system, a film develops from the ideas of individual directors who make their pitch to producers. Then the script is written by the director and a writer or team of writers. Sometimes no real screenplay exists or, as in the famous example of Wong Kar-wei, the director writes the screenplay during the shooting process. Thus Hong Kong films are characterized by their episodic nature, which Bordwell traces back to the influence of martial arts and the Peking Opera (183–5), maintaining that popular mass entertainment stages "the tension between 'spectacle' and 'narrative'" (178). Episodic narratives lend themselves well to the depiction of urbanity, characterized by chance encounters, disjointed experiences, and alienation as theorized by Georg Simmel and Walter Benjamin, with whom we are already long familiar.

Urbanism in Hong Kong cinema

Leung Ping-kwan points out that urban culture centrally defines Hong Kong's identity and differentiates it from mainland China. In fact, Hong Kong has developed much of its culture specifically in historical dialogue with China – witness two key dates: 1949, the founding of the People's Republic of China (PRC), and 1966–76, the years of the Cultural Revolution. Ping-kwan explains that the "mainstream literature and cinema in 1930s China," was characterized by "a clear-cut dichotomy between the city and the country" in which the city embodied "temptation, corruption, vice, and cunning manipulation," and the country "innocence, uprightness, and fraternity" (228). Not surprisingly, 1930s Chinese cinema tended to favor the country over the city.

In the 1950s, the choice of the refugees to either stay in Hong Kong or return to China was played out in the Hong Kong cinema – as a "negative depiction of Hong Kong's urban space with particular emphasis on its poor living conditions and the avarice and selfishness of the residents in a capitalist society," or as a utopian representation in which Hong Kong was represented as a "lawful, just, rational, and dynamic place where diverse attitudes could be accepted" (228). A third model was a "satirical comedy to represent funny and sympathetic individuals who seek survival in the commercial world" (231).

Hong Kong arrived at a turning point during the unrest of 1967 against British colonial rule, when pro-communists were inspired by the Cultural Revolution in the PRC and organized large-scale demonstrations, strikes, and riots in the city. Afterwards, the government organized such events as "the Hong Kong Festival, pop parties, fashion shows, the Miss Hong Kong Pageant and so on, to design a modern, Westernized image for the people of Hong Kong, in order to make the residents of the colony identify less with its mother country" (Ping-kwan 233). The validation of western values ultimately led to a balance of western and Chinese cultural values in Hong Kong cinema (235). Contemporary urban popular culture was also influenced by television (TVB), which began in 1967 and offered popular series "with an urban background" that led to the creation of "Canto Pop," Cantonese pop music (236). Most important for the Hong Kong New Wave and the reflection of the urban environment of Hong Kong in films, however, was the fact that many of the new generation of filmmakers were born in Hong Kong but trained as directors abroad. Their "self-awareness of the city and its representation" was reflected in literary texts and films that employed "double or multiple perspectives in their narratives to examine Hong Kong's urban space" (238–9), for example in Tsui Hark's *Dangerous Encounter of the First Kind* (1980), Allen Fong's *Father and Son* (1981), and Ann Hui's *The Secret* (*Feng jie*, 1979).

During the years leading up to the important moment of the return of Hong Kong by Britain to China in 1997, the cinematic representation of Hong Kong took on the function of an allegory, which Ping-kwan illustrates with the examples *The Boat People* (1982), by Ann Hui, and *The Wicked City* (1992), produced by Tsui Hark and directed by Mak Tai-wai. He also points to the nostalgia in such films as Stanley Kwan's *Rouge* (1987), favored particularly by Hong Kong people before 1997 because "the moment of uncertainty and anxiety" made them "desperate in their search for an identity" (244). A more realistic portrayal of Hong Kong, according to Ping-kwan, was offered by such films as Ann Hui's *Summer Snow* (1995) and post-1997 films that reflected "the economic recession that occurred after the handover" and subsequently showed "marginal and alternative spaces of Hong Kong," such as "the gay community, the youth in the poor housing estates, the prostitutes from the north" (249). Examples include Stanley Kwan's *Hold me Tight* (1998), Fruit Chan's *Made in Hong Kong* (1998), and Yu lik-wai's *Love Will Tear us Apart* (1999).

The significance of urbanism for Hong Kong film is limited neither to the ideological dispute between Hong Kong and mainland China nor to the economic development of Hong Kong, because its film industry also could rely on a special relationship with its high-density urban audience. Because of the generally cramped accommodations, those living in Hong Kong go out in the evening and prefer collective but anonymous film screenings over video or cable-TV in their crowded homes. In the 1950s, theaters were neighborhood centers and inhabitants of Hong Kong became used to seeing their city portrayed on the screen (Bordwell 36). Bordwell suggests that the rapidity of the dialogue and action mirror the speed of daily life in Hong Kong. He focuses on the tradition of midnight screenings, which reflects the urban practice of film reception, and which created a particular local market and a unique instant-feedback system between film-makers and audiences that enabled the Hong Kong film industry to be a readily responsive one.

Martial arts cinema

Paradoxically, martial arts cinema is not a particularly urban genre, but it emerged out of Hong Kong and constituted the foundation of its film industry. As a genre, it embodies the contradictions of modern, urban, industrial society and the diasporic projection into the past. Hong Kong action films have used martial arts and the history of China to create a mythical past of a lost homeland. As the most famous genre associated with Hong Kong film, martial arts cinema represents a crucial counterpoint to and predecessor of the urban cult of the heroes cycle, which I return to below. Martial arts cinema created its own film language, its own cult audience, its own production companies, its own networks of circulation, and its own stars.

Bordwell suggests that martial arts films reflect the historical development of martial arts themselves, and in fact martial arts films did not rely on choreography for fight scenes until the 1950s and 1960s, when editing became more rapid (206). Lent explains that in the swordplay films of the 1950s and early 1960s the Confucian code dominated, with plots revolving around "filial ties, destruction of which led to violence and revenge, and the master–pupil relationship" (115). In the 1970s martial arts films adopted the more action-oriented kung fu, hence the phenomenon of Bruce Lee, star of Wei Lo's *Fist of Fury* (1972), *Way of the Dragon* (1972), which he directed himself, and *Enter the Dragon* (Robert Clouse, 1973).

Lee was born in San Francisco's Chinatown while his parents were touring the US with a Cantonese opera troupe and waiting to receive American citizenship. In 1941 the family headed back to Hong Kong, where his father worked in the film industry. In the popular imagination, Bruce Lee's biography is that of the émigré who returned to his homeland and became "the territory's most famous citizen," celebrating "Hong Kong identity" (Bordwell 50). So the kung-fu martial arts films produced in Hong Kong in the 1970s "hastened the end of the didactic, tradition-laden Cantonese cinema" and laid the "foundations of the New Wave and the slicker films of the 1980s" (Bordwell 207). While Bruce Lee is without doubt the most important martial arts star of the Hong Kong film industry – his global fame has exceeded his lifetime – Jackie Chan will be the focus here, because I read him as a star successor to Bruce Lee who was intentionally created by the Hong Kong martial arts film industry and who functions as a transitional figure from a diasporic concept of traditional martial arts to a transnational star associated with deterritorialized urbanity.

Jackie Chan, star of transnational urbanity

Whereas Bruce Lee is associated with the traditional martial arts and Chinese diaspora, Jackie Chan represents and functions in a network of urban transnationalism. He is one of the most important popular and mainstream stars coming out of Hong Kong cinema today. His star persona and his roles are symptomatic of the tension between the idealized and mythologized past and the pragmatic, urban attitudes of Hong Kong culture. Chan's rise to stardom also symbolizes the successful transnational exchange between Hong Kong and Hollywood, since his films have been distributed in the US and it is from there that he has become an international star.

In order to function as a global commodity, however, Jackie Chan's star persona necessarily diffuses the specific reference to Hong Kong identity. Kwai-Cheung Lo argues: "People in the Hong Kong film industry are therefore being presented with the opportunity to compete in Hollywood at the same time that they are losing the battle of being able to compete with Hollywood in their Asian markets" (128). In contrast to traditional European national cinemas, Hong Kong cinema never was opposed to Hollywood, but rather adopted a positive relationship to it. Chan's transition from what is coded as East to what is coded as West was apparently without difficulty, particularly in the following films: Woo-ping Yuen's *Drunken Master* (1978), Hdeng Tsu's *Rumble in Hong Kong* (1974), Stanley Tong's *Rumble in the Bronx* (1995) and Brett Ratner's *Rush Hour* (1998). Bordwell captures Jackie Chan's star persona in relationship to Hong Kong: "The star's innocent, indomitable urge to overcome all obstacles seems designed to project one modern image of Hong Kong"; Chan embodied "calculated cosmopolitanism," and by the 1990s he was "an emblem of Hong Kong itself" (58–9).

Chan was born in Hong Kong in 1954 and then moved to Australia with his parents, but then the family returned to Hong Kong for Chan to study Peking opera at the China Drama Academy, so he experienced the traditional training of the opera school and the urban environment of Hong Kong simultaneously (see Fore 127). The generic conventions of kung-fu comedy, particularly as practiced by Chan, center on a rather haphazard student of martial arts whose family, or *dojo*, is confronted with a problem, which he ultimately resolves through fighting – after he has taken some severe beatings by his adversaries.

Jackie Chan is associated with transnational urbanism – films show him in Hong Kong, New York, Los Angeles – even though he has acted in several traditional Hong Kong action films, portraying a hero in a Shaolin temple. The transnational public is primarily familiar with his Hollywood comedies, but he also has a cult following around the world that celebrates his entire oeuvre. His characters integrate acrobatic martial arts with comedic elements; he often plays the naïve innocent, sometimes even the victim, with whom the audience can identify, but he generally wins in the end, often against a whole group of adversaries. Since he achieved international stardom in the binary moral logic of action films of good and bad characters, he has been cast as a *good* character, even though the re-released early Hong Kong film *Rumble in Hong Kong* shows him as the leader of a criminal gang.

Steve Fore contextualizes Chan's shift from a regional to a transnational star in his particular path of migration which occupies "a cultural space between the involuntary migrations of political and economic refugees and the voluntary cross-border travels of expatriate professionals employed by transnational corporations" (116). Fore explains that in the context of globalization, Chan's star persona changed from relying on core values associated with the Hong Kong community to becoming increasingly "more diffuse and less emphatically tied to a specific cultural space" (117). Like many scholars, Fore situates Chan in relation to Bruce Lee, whom he sees as "the most potent Hong Kong-specific symbol of cultural China and of a particular version of the Chinese nation and Chinese nationalism" (118). In contrast, he suggests, Jackie Chan represents entertainment. He bases this difference on a conscious differentiation of Chan's from Lee's star persona, on the one hand, and on the different socio-political contexts of the early 1970s, the time of Lee's death, and the late 1990s, the time of Chan's rise to global stardom, on the other.

According to Fore, "Kung-fu comedy," the subgenre created by and specifically associated with Jackie Chan, marked the "transformation of Hong Kong from a colonial backwater to a rapidly modernizing and fast-paced urban capitalist society," because these films combine "pragmatism, cynicism, personal ambition, rebelliousness, ruthlessness, acquisitiveness, and quick-wittedness," a set of values quite different from the one that underlies traditional martial arts cinema (126). These characteristics were communicated readily to a new generation of Hong Kong inhabitants, who were separated from China and increasingly aligned themselves with western notions of urbanism. Kwai-Cheung Lo argues that before Chan found a distributor to launch the first US nationwide release of *Rumble in the Bronx* in 1996, which became a hit, the Hong Kong action film had only limited exposure in the US – in "Chinatown, inner-city theaters, second-run houses, small-town double-bills, and drive-ins" (131). Bruce Lee's films circulated in urban B-movie houses, particularly among minority audiences, and also represented resistance among disenfranchised young men globally. Jackie Chan's films aimed at the mainstream of global circulation, but were narratively and topically situated in transnational urban spaces.

Before his global success, Chan's characters were defined by the tension "between tradition and modernity in the context of present-day Hong Kong" (Fore 132). In order to become a "global product," however, his films shifted from martial arts to stunts and props, from a Hong Kong identity to a vague understanding of being Chinese (Fore 132; also Morley and Robbins 109, 113). Lo makes a similar point when he explains that in Hollywood films characters from Hong Kong appear as "mainland Chinese," which negates "Hong Kong's particularity" (129), even though the American mainstream press points out that the individual actors and directors are indeed from Hong Kong. He concludes that "the subject called 'Hong Kong' by the western media is reserved primarily for real, external portrayals at the same time that it is virtually excluded from the fictional, diegetic world" (129). Because Hong Kong stars in Hollywood generally act as characters portrayed as "generic[ally] Chinese," those films rely on and reinforce American stereotypes about the Chinese (133).

Stanley Tong's *Rumble in the Bronx* (1995) very consciously situates Chan in an American context. His character, Keung, arrives confused and overwhelmed at the airport, where his uncle picks him up. His English is imperfect and his awe of Manhattan portrays a naïveté projected onto a stereotypical Chinese immigrant that stands in stark contrast to the urban sophistication of Hong Kongers. Keung stays with his uncle in the Bronx, where a motorcycle gang terrorizes the neighborhood, including his uncle's store. The assimilated uncle leaves Keung behind to guard the store. At night the motorcycle gang meets under Keung's window and puts on a show of destruction and aggression. Keung, dressed in an infantilizing one-piece jumpsuit, takes to the street and attempts to protect the car that is in his care. The dystopia of inner-city violence is accompanied by a mix of markers of American subculture: the gang's exaggerated urban clothes, their leather outfits, the graffiti on the walls, and the motorcycle culture, all of which was almost outdated in the mid-1990s. In contrast to the urban threat, Keung is dressed in nice clothes, clean-shaven, helpful, and yet partially helpless. His figure is offered for identification by all those who feel bullied. Bordwell reads Jackie Chan as an inversion of Bruce Lee, a "passionate masochist" (58). Thus, Keung gets beaten up repeatedly until he saves the day, reinforced by his sidekick, a little boy in a wheelchair, whose attractive sister is involved with the gang leader.

In order to fight or escape the gang, Keung has to negotiate the urban jungle: running through a parking lot, jumping over railings, climbing up the outside walls of a building, and jumping across the street from a parking garage to the balcony of another building. Chan has made a name for himself by performing his own stunts and turning an action film into fantastic acrobatics. *Rumble in the Bronx* employs some of the features of Hong Kong action films in that jumps from the parking garage to the balcony of the next building are shown from three angles, a technique that breaks the illusion of the film and foregrounds the fact that we are watching martial artist–acrobat Jackie Chan performing a stunt and not just the character Keung escaping from his torturers. Fore explains:

> The payoff shot of these stunts is always filmed in a single take (to confirm their authenticity) and with multiple cameras. The resulting action is edited into the finished film in a rapid sequence of the same action shot from three or four different perspectives (highly reminiscent of the sailor smashing the dishes in *Battleship Potemkin*). In almost all of his films since *Project A*, Chan has ended his film with the credits scrolling over outtakes from the film we have just watched. In these outtakes, actors breaking down in laughter over blown lines alternate with shots of carefully arranged stunts and martial art moves going badly wrong.
>
> (131)

These cinematic strategies harken back to the body politic of Hong Kong martial arts cinema and situate the reality of the body and its action in a new context of urban America. By using generic urban structures, such as parking garages, staircases, and skyscraper walls as the terrain for Chan's acrobatic action style, the cityscape stands in for any city, anywhere. Instead of an heroic celebration, we find an ironic and self-deprecating performance of the physical skills that result from discipline. Instead of conquering an enemy, defending honor, or standing up to an oppressor, physical prowess enables the main character to immerse himself in urban environments and fend off urban threats.

Kwai-Cheung Lo, as well as Amy A. Ongiri, emphasizes the subcultural cult circulation of Hong Kong action films among American minorities, particularly during the 1970s, around the heroic figure of Bruce Lee, who embodied resistance to the colonial powers that could be appropriated for other forms of resistance. They imply that the contemporary mainstreaming of the Hong Kong action film in Hollywood as exemplified by the star persona of Jackie Chan represents a depoliticization of the genre. Laleen Jayamanne, however, points out that Chan's four successful films in the US – *Rumble in the Bronx*, *Who Am I?* (Benny Chan Muk-sing and Jackie Chan, 1998), *Rush Hour* (Brett Ratner, 1998), and *Rush Hour 2* (Brett Ratner, 2001) – "have either an African or African-American connection" (151). She reads these configurations of ethnic characters (also including *Shanghai Noon* [Tom Dey, 2000]) as "a transnational kinship group by creating filial networks with familiar generic types" (155). These kinds of "transnational kinship group[s]" presuppose the urban space in a transnational cinematic culture; they do not advance resistant, heroic fight against oppression modeled by the earlier star and identification figure of Bruce Lee.

The heroes cycle: urban cool

None of the martial arts films by Bruce Lee or Jackie Chan were explicitly marketed as urban films. But in the late 1980s, Hong Kong cinema exploded with yet another hybrid genre that was fully reliant on the depiction of urban cool. John Woo's *A Better Tomorrow* (1986) began a series of gangster films that came to be known as the "heroes cycle." It integrated the setting of modern Hong Kong, the American tradition of the gangster film, including the excessive use of gun violence, traces of the martial arts tradition, and a cool, urban aesthetics. These new urban thrillers offered stylized, sometimes hyperviolent, male melodrama centered on a hero, but they also, according to Jinsoo An, worked through the anxiety of Hong Kong's impending return to China, attracting audiences with the "relationship between Woo's exuberant cinematic style and the social anxiety driven by the historical situation of Hong Kong" (95). An also suggests that by situating Woo's films in the context of the national cinema of Hong Kong, critics attempt to rescue them from the lower status of "exploitation action flicks or cult movies." He considers Woo's some of the most important cult films defined by a particular fan base in the US, but he also regards them as representing "[t]ransgression, or the violation of boundaries" (97). Woo's films share the setting of postmodern urbanism, masculine action, and male melodrama based on male friendship and loyalty coded as heterosexual by the inclusion of a female victim who is treated with honor and respect. The heroes cycle integrates the local tradition of kung-fu martial arts films and setting of Hong Kong with the transnational conventions of the gangster film.

In Woo's *Hard-Boiled* (1992), jazz music from the 1970s connects the contemporary urban environment of Hong Kong to American 1970s gangster and exploitation films and American television series such as *Kojak* and *The Streets of San Francisco*. The centrality of jazz also connects the heroes cycle to film noir, in which jazz connoted the marginality of African-American urban culture and presumed sexual transgression. Hong Kong action film was received in the urban milieu of 1970s American film culture when kung-fu films were screened in inner-city cinemas, developed a cult following among African-Americans, and were part of an exchange with Blaxploitation with regard to topics, narrative, production, and actors (see Ongiri 2005a, 2005b; also Prashad for the relationship of African-Americans and Hong Kong action cinema). By turning the main character of *Hard-Boiled* – Tequila, played by Chow Yung Fat – into a jazz musician, the film creates an intertext for cult fans.

The opening of *Hard-Boiled* creates a postmodern pastiche of references that not only is emblematic of Hong Kong but also posits a close affinity between contemporary urbanism and postmodernism. The film begins with a close-up of a drink over which the credits are projected, and then we hear jazz music with a 1970s theme. The music continues through several shots of Hong Kong with different street scenes until the narrative begins in the traditional Wyndham Teahouse, in which cages containing birds are sitting on tables and men are drinking tea. Woo employs fetishizing slow motion to create suspense when violence is about to ensue. We hear fragments of conversations, one concerning emigration, when one patron of the Teahouse asks another whether he has ever considered emigrating and receives the answer "No, this is my home." The conversation continues: "There are Chinese restaurants abroad," and the patron answers, "But we have the original."

This minor dialogue touching on questions of home and migration, authentic Chinese food and the Chinese diaspora frames the upcoming violence and contrasts the seemingly idyllic Teahouse with the fact that it is a front for illegal gun-trading. In the violence that erupts the Teahouse is shot to pieces by gangsters wielding an assortment of guns. Sudden violence is the mark of the urban gangster film portrayed as a total shoot-out seen from various camera angles and using freeze frames and slow motion. The interior violence is framed by shots emphasizing exterior, neon advertisements associated with cinematic representation of the urban space since Weimar cinema.

The toughness of men is one main trope for gangster films associated with urbanism. A friend remarks on how tough Tequila is now, implying that he was different in the past, which is reflected in his jazz artistry. This allusion to an innocent, past persona is part of the appeal of the characters played by Chow Yung Fat: a funny, likable innocent man, forced to be an undercover cop and unable to get the girl in *City on Fire*; a displaced jazz musician as cop in

Hard-Boiled; a man trying to save the woman he blinded in *The Killer*; or the sidekick who gets humiliated in *A Better Tomorrow*. These examples of the kindness, silliness, and tragic imperfection of Yung Fat's characters contradict the dominant academic view, according to which the films of the heroes cycle portray only a static and tough masculinity. There is clearly always a tinge of nostalgia for a different, past masculinity that, tragically, has had to be sacrificed for this new urban cool.

A Better Tomorrow demonstrates the urban "cool" look, which continued throughout the heroes series, including elegant long coats, sunglasses, and cigarettes in a subtle echo of film noir. It also emphasizes surveillance, television, and the high-tech industry of urban crime, differing from the moral stories of the martial arts films that centered around colonialism, honor, and transgression of morality. *A Better Tomorrow* connects the experience of the urban environment and mobility with the transnational power of cultural imperialism, the dollar. The film's opening vignette famously shows the characters played by Leslie Cheung and Chow Yung Fat lighting their cigarettes with forged dollar bills. Once the credits have rolled and we have been introduced to the technology used for forging the foreign currency, we travel by car through the city, seen from the perspective of the car tires, connecting the circulation of illegal currency with the circulation of the gangsters through the city.

The heroes cycle is characterized by two forms of doubling: firstly, films often offer two endings, one in which the narrative comes to a tragic conclusion and the hero dies violently, and one that seems to be located in a fantasy or memory, in which the character is still alive; and secondly, the configuration of the male hero is often doubled through a senior colleague, a substitute father, a good friend, or a partner in crime, in addition to the larger male bond of the gang. Bordwell claims that it is the poverty of the narrative that leads to so many narrative doublings in Hong Kong action cinema (185–6), but I suggest that the structure of male melodrama also demands the doubling of the main male character. The context for male melodrama is the public space, whereas female melodrama takes place in the private space of domesticity, most commonly associated with the mother–daughter dyad, which is entirely absent in this genre. Women are either hurt or incapable of understanding, hence tragically unable to support the men, while a symbolic father–son, older–younger friend, mentor–mentee configuration provides the basic structure for the male–male relationships, warding off any suspicion of homosexuality and constituting an unacknowledged aspect of the urban heroes cycle.

The New Wave

The urban and cosmopolitan filmmakers of the so-called New Wave belong to the generation born after the war, who grew up without memories of mainland China and who were often educated abroad, raised on popular culture, and trained in television. The New Wave emerged in the early 1990s and includes filmmakers Yim Ho, Allen Fong, Tsui Hark, Stanley Kwan, Ann Hui, and Wong Kar-wei. The movement was furthered by cultural funding when in the mid-1990s the Hong Kong Arts Development Council made money and production and post-production facilities available for young directors to create film and video shorts (see Bordwell 262). And according to Lent, at the same time that the directors of the New Wave "revealed the myth of urban prosperity, the dissatisfaction of youth, the uncertainty about Hong Kong's future and identity, and the myriad problems and societal changes of the Crown Colony," they also created striking new images of the city with fragmented and sometimes mysterious narratives in beautiful shots and succulent colors (111). This was a new urban cool that had an intimate and mysterious quality to it, more evocative of the French New Wave than the Hong Kong action film. Wong Kar-wei's famous *Chungking Express* (1994), for example, creates a combination of detachment from and affection for the urban neighborhood similar to the films discussed in Chapter 3, now based on postmodern confusion, loss, melancholia, and missed encounters and misunderstandings. Wong Kar-wei, one of the most famous of the directors to emerge from the group, is described by Bordwell "as the allegorist of postmodern urban culture and Hong Kong's pre-1997" anomaly (270).

Wong Kar-wei is seen as one of the filmmakers able enough to capture the postmodern qualities of contemporary cities such as Hong Kong. Ping-kwan posits that in *Chungking Express* Wong Kar-wei creates a postmodern pastiche out of different parts of the city, but because the names refer to real places they reconstitute the cinematic city from its parts –

> a postmodern pastiche of Chungking Mansion in Tsimshatsui and a fast-food place called Midnight Express in the Lan Kwai Fong area in Central. The pastiche of the names of places in the title, like the pastiche of the two unrelated stories in the film, helps us to blur geographical divisions and discredit referentiality. Yet the use of the actual names of these two places, as well as the sensitive lingering of the camera and the attention to details in art direction, also redirect our attention to the specific urban sites in Hong Kong.
>
> (245)

In this understanding of postmodernism, the cinematic representation is attached to the urban reality but not bound by it, and ultimately reflects and heightens the fact that the urban space itself is already postmodern.

Wong Kar-wei's films portray postmodern urbanism not only in regard to Hong Kong, but also in a transnational urban context. His film *Happy Together* (1997) portrays two male lovers from Hong Kong in the beautiful and displaced space of Argentina, traveling between the famous waterfall Foz de Iquazu and Buenos Aires. The narrative structure is episodic, portraying their arguments, their attraction to each other, and their alienation from each other and their surroundings. The soundtrack consists of a tango by famous Argentinean musician Astor Piazzolla. One of the most striking scenes in the film shows the two lovers dancing the tango in the communal kitchen. The stark setting consists of all-around tiled floors and walls, and captures an alienating, cold quality in contrast to the erotic warmth associated with the tango. At the same time, the space is so obviously artificial and beautifully aestheticized, bathed in blue, that in combination with the music it captures the tango's sense of longing. The dance is performed by a homosexual couple expressing both desire and impossible hope through their body language, which is less upright and precise than the movements that traditionally characterize the tango. Instead, the characters follow the steps but interpret the dance in ways that connote desire, despair, and loss.

One indicator of the intimate relationship between Wong's films and a transnational understanding of urban space is the fact that, as Bordwell points out, the releases of the films in Japan are accompanied by "a souvenir program, often including maps of Hong Kong pointing out where each scene takes place. For *Happy Together* Wong's Japanese partners produced an *objet d'art* chapbook consisting of collages, scraps of text, and a few mementos" (271). That mementos of a film set in Argentina and produced in Hong Kong are distributed to an audience in Japan illustrates the function of memory and nostalgic attachment in a deterritorialized and postmodern world.

In conclusion, we find in these different developments – from heroic Bruce Lee to self-deprecating Jackie Chan, from mythological but distinct past to exchangeable transnational present, from distinct genres to hybrid appropriations of generic conventions – the changes that reflect larger shifts from national to transnational and from modern to postmodern cities and cinema.

Case Study 4 **Ringo Lam's *City on Fire* (1987)**

"If Woo films his dreams, Lam shoots his nightmares." This is how Lisa Odham Stokes and Michael Hoover summarize the difference between John Woo and

Ringo Lam, the two most influential filmmakers of the heroes cycle (64). *City on Fire* was released six months after *A Better Tomorrow*, and both are set in urban Hong Kong and are variations on the gangster film. Lam's background in television and his study of film production at York University, in Toronto, typifies his generation. After his return to Hong Kong in the early 1980s, he first directed comedies. *City on Fire*, the first film that he wrote *and* directed, began the series, which continued with *Prison on Fire 1* (1987), *School on Fire* (1988), and *Prison on Fire 2* (1991). These films were responses to the uncertainty of Hong Kong's future and an increase in violent crime (see also Stokes and Hoover 65).

The main character in *City on Fire*, Chow, played by Chow Yung Fat, is an undercover police officer who in his last operation befriended a criminal named Shing. Chow considers himself responsible for Shing's death and is tormented by nightmares about it. His supervisor, Inspector Lau, makes careless decisions that cost the lives of his undercover agents, and he forces Chow to take up the position of his former undercover agent, Wah, who has been killed. The narrative is motivated by Chow's conflict between loyalty and duty: loyalty to his friends (even if they are gangsters) and his girlfriend Hung, and his duty as a police officer. Because Chow is undercover, not all the members of the police force know that he is one of them, and he is chased through the city, arrested, and beaten. After the gang he infiltrates robs a jewelry store, the members are hunted down by the police and surrounded, and Chow is suspected by the gang leader of being an informer. In the final scene, most of the members of the gang and Chow are shot, and the film concludes with the double ending of Chow's tragic death and a shot of him dancing in the streets when he was still alive.

The urban topography that characterizes *City on Fire*'s cinematic depiction of Hong Kong consists of transitional spaces, such as bars and hotel rooms, and the cityscape through which Chow is hunted and that increasingly closes in on him. He is introduced as he enters a bar, a privileged setting in the topography of city films. Later, when he is undercover, Chow stops over in a run-down hotel room, a transitional space, which marks his lack of a real home. The narrative development of the film shows the police closing in on him since he has no space to retreat to. Because Chow is an undercover agent he is tragically caught between the two sides of the law. Near the beginning Chow is arrested and taken to a police station where we see another element of the urban setting: the morgue. Chow reflects Hong Kong's fate, determined by others, of which neither the old nor the new presents a good choice. Hong Kong becomes a trap, like the small apartment in which Chow is first holed-up with the gangsters. A shot of the

continued

gangster Fu and Chow shows their intimate friendship, and in a repeat of the nightmare Chow now sees Fu's face instead of Shing's, awake and looking at him, foreshadowing his guilt.

The centrality of Chow's nightmare for the narrative is reminiscent of film noir on the one hand and older Chinese-inflected narrative traditions on the other. In film noir, dreams refer to the unconscious and reflect a Freudian understanding of the psyche. In the Chinese culture of ancestor worship and parallel ghost worlds, ghosts can carry different meanings and are the topic of an entire subgenre of the Hong Kong action film. The motif refers to these different contexts simultaneously and shows the hybrid quality of the text in question. The guilty memory of Shing's death produces Chow's recurring nightmares, which are shot in blue tones and slow motion, and show policemen with flashlights walking around an unknown building until we see the body of a man who has been shot, his blood splattered on a window. The dead man then opens his eyes and blames Chow for his death, upon which Chow rises out of his sleep and calls out to the dead Shing. The dream itself is surreal, as is Chow's waking scene, in an undeveloped *mise-en-scène*. Extreme shadows from window blinds mark Chow's body, giving the film a symbolic quality and again referencing the aesthetics of film noir. After Chow is beaten by the police, and can hardly walk because of his injuries, he is shown in a scene shot entirely in blue, from the perspective of a car looking through the rain-swept windscreen, wipers moving. The color coding of these shots connects the undeserved punishment of Chow with his dream life and his feelings of guilt over his friend Shing's death.

The music and coloring of the film create a tonal and aesthetic quality that ties *City on Fire* to postwar culture in the US in general and African-American culture in particular. The colors blue or red are emphasized, and the music is reminiscent of jazz and the blues. Like Woo's *Hard-Boiled*, *City on Fire* begins with jazz, a saxophone solo accompanying the credits, in blue on a black screen. When a synthesized 1950s-themed melody is added, the camera pans down into the streets, emphasizing the neon signs, similar to the opening of the noir film *Criss Cross* (discussed in Chapter 2). Once the camera arrives at the street level, however, instead of the parking lot of a bar, as in *Criss Cross*, we find ourselves in a bustling marketplace. The camera shifts from the establishing shot to the *cinéma vérité* style of the hustle and bustle of the Asian market – as Stokes and Hoover point out, "with live shooting and tracking camera," which emphasize the "realism" of Lam's portrayal of Hong Kong (65).

The emphasis on loyalty and honor connects the heroes cycle to the earlier martial arts films, recast in the modern setting of the city and in the gangster genre. The question of male friendship and betrayal is central to male melodrama: Chow is haunted by the trauma of betrayal and in the end sacrifices himself to keep from repeating it. He is put in impossible positions, unable to fulfill what is expected of tough masculinity, betrayed by the father figure, suffering from his guilt, unable to keep his girlfriend Hung, and unable to function authentically in either male bond, that of the police or that of criminality. Like many films in the heroes cycle, this one celebrates the style of fetishized hypermasculinity with shots that present the toughness of men in contrast to Chow's tragic lack, which, however, is endearing when the film repeatedly shows the scene of him dancing in the street.

The film ends with a melancholy, masochistic excess of destruction bathed in blue tones. The killing of the gangsters, including Chow, takes place at a foggy and dilapidated industrial building that is aestheticized to the degree that it functions as a symbolic space. Only Chow and three gangsters are left; Chow is injured and the gang's boss realizes he is an undercover agent because of the police presence. Chow is drenched in blue on the bed by the shadows of the bars on the window. The sky repeats the motif of blue, and inside shadows cast all of them in blue tones. They are trapped. Light and fog come in through the bullet holes in the walls. The blue coloring and the saxophone soundtrack create that urban blues feeling associated earlier with Chow's sense of guilt. Fu tries to flee with Chow from the police bullets, but Chow confesses that he is a policeman and has betrayed him. He tries to sacrifice himself for his friend. The final shot shows Chow on the floor drenched in blue, and after a cut we see him dancing in his blue clothes and his dead body in a double exposure in what Bordwell calls the double ending.

City on Fire is a highly stylized film, relying heavily on aesthetics created by color, especially blue, which creates the "cool" urban look. The film integrates the traditional topic of honor into a modernist setting of urban crime. As in others of this genre, the film is structured around fetishized masculinity which, however, negotiates between duty, desire, tradition, and loss, much like Hong Kong at the time.

Further reading

David Bordwell (2000) *Planet Hong Kong: Popular Cinema and the Art of Entertainment*, Cambridge, MA: Harvard University Press. This book provides an overview of Hong Kong cinema in the context of film studies.

Poshek Fu and David Desser (eds) (2000) *The Cinema of Hong Kong: History, Arts, Identity*, Cambridge: Cambridge University Press. This collection addresses Hong Kong cinema within a broad historical range, including essays that focus particularly on the geographical and political aspects of Hong Kong.

Meaghan Morris, Siu Leung Li, and Stephen Chan Ching-kiu (eds) (2005) *Hong Kong Connections: Transnational Imagination in Action Cinema*, Durham, NC: Duke University Press. The edited collection offers particularly interesting case studies of transnational cinematic exchange organized around Hong Kong action cinema.

Lisa Odham Stokes and Michael Hoover (eds) (1999) *City on Fire: Hong Kong Cinema*, London: Verso. *City on Fire* provides the perspective of Stokes and Hoover on Hong Kong cinema with a particular emphasis on contemporary cinema, addressing but subordinating the history of Hong Kong action cinema to its contemporary cinema.

Esther C. M. Yau (ed.) (2001) *At Full Speed: Hong Kong Cinema in a Borderless World*, Minneapolis: University of Minnesota Press. This is an important collection that brings together scholars of Hong Kong cinema. The essays address the New Wave, martial arts, and the politics of place in relationship to transnationalism and Hong Kong cinema.

Essential viewing

Robert Clouse. *Enter the Dragon* (1973)

Wong Kar-wei. *Chungking Express* (1994)

Ringo Lam. *City on Fire* (1987)

Stanley Tong. *Rumble in the Bronx* (1995)

John Woo. *A Better Tomorrow* (1986)

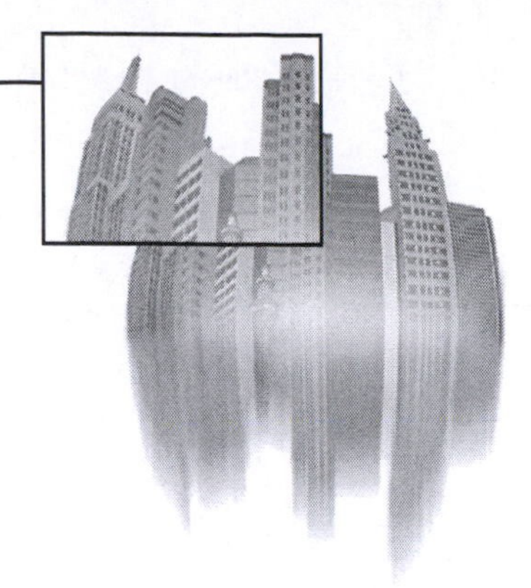

5 The city in ruins and the divided city: Berlin, Belfast, and Beirut

The fronts are everywhere. The trenches are dug in the towns and the streets.
Winston Churchill

Learning objectives

- **To understand the two different traditions and functions of the trope of the ruin**
- **To describe rubble films and engage with the moral, historical, and political questions they raise**
- **To contextualize the postmodern use of the ruin in its historical, geographical, and philosophical dimensions**
- **To be able to analyze the spatial politics of films about war, resistance, and divided cities as reflections of ideological positions**

Introduction

This chapter discusses the cinematic representation of the destroyed, the ruined, and the divided city and relates films about war and postwar moments to the conditions for production in destroyed, occupied, and divided film industries. In order to capture and put analytical pressure on the relationship of historical conditions of destruction and their aesthetic manifestations, the chapter is organized around the figure of the ruin on the one hand and the spatial topography of the divided city on the other. Ruins can have two different functions, which are rooted in distinct traditions: on the one hand, they mark precise historical moments, for example in the rubble film of the immediate German postwar moment. In these films from 1946–48, Berlin in ruins becomes the site for negotiating guilt, redemption, and rebuilding in regard to the Holocaust and the Second World War. On the other hand, ruins as a postmodern cipher invoke historical moments and iconic images but empty them of their historical and geographical specificity in what I call the retro-rubble film.[10] Marc Caro and Jean-Pierre Jeunet's *Delicatessen* (1991) and Lars van Trier's *Zentropa* (1991) are postmodern fantasy reworkings of the city in ruins that conjure up iconic images of war-torn urbanity but without being bound by historical accuracy. The latter half of this chapter deals with films about divided cities, which in contrast to rubble films do not constitute a cycle. Instead, the topography of division relates to films set in a relatively small number of cities that represent political and historical anomalies, and here the chapter focuses on films about Berlin, Beirut, and Belfast.

Cities and war in urban studies

"Warfare, like everything else, is being *urbanized*," explains Stephen Graham, and he goes on to show that "cities are key sites [in the] 'new' wars" that are being fought in the post-Cold War era (2004a: 4). This development began during the Second World War when the conflict moved from clearly demarcated battlefields to the urban environment, where it affected life in the city and the daily experience of its citizens. Michael North, for example, points out that by May of 1941 one-sixth of Londoners had been made homeless (437). Whereas the First World War is associated with the names of battlefields, the Second is marked by a list of cities synonymous with absolute destruction. However, Graham sees another contemporary shift in the post-Cold War and post-9/11 periods that largely "*entail systematic and planned targeting of cities and urban places*," and he suggests the term "'urbicide': the deliberate denial, or killing, of the city" for "the intersections of war, terrorism, and subnational – specifically urban – spaces" (2004a: 24–5, italics in the original).

Graham discusses industrial cities of the global North in the nineteenth and the twentieth century that had caused warfare but also constituted prime targets because of their industries (2004a: 2). In the current development of globalization, cities are part of transnational networks of violence, and Graham connects the destructiveness of war in urban environments to the violence surrounding the oppression and resistance of minorities in the city: his examples of urban warfare include the "massive onslaughts by US and British forces on Basra, Baghdad, Fallujah, Kandahar, Kabul," as well as "the LA riots of 1992," the "continuing suicide bombings in Israeli bars, buses, and malls," and the "resource- or drug-fueled guerrilla wars in Freetown, Bogotá, and Monrovia" (2004a: 5). Graham suggests a continuum of different forms of violence within urban spaces and points, for example, to the militarization of "systems of criminal justice, law enforcement, and public space regulation" (17). According to Graham, "[b]ecause both the 'homeland' and most colonized and invaded spaces are becoming more and more urban, urban terrain increasingly provides the 'battlespace' for the US military in both spheres" (18).

Graham's argument reflects contemporary wars, especially those between the advanced military forces of the US and their targeted cities in poor, Islamic countries. Graham's observations about current warfare and cities might be better applied to contemporary new media, such as computer games, often developed in the military–industrial complex and then transferred into the entertainment industry. The examples of the traditional medium of film discussed in this chapter reflect historically earlier periods but attests to the increasing interconnection between cities and wars.

Divided cities pose a different and particular challenge to urban planning and show exceptional characteristics in urban development; however, contemporary approaches to the unique situations of Berlin and Belfast shift to an emphasis on the potential for realizing productive and culturally inclusive visions of integration (see Neill and Schwedler). "Wounded cities" is a concept by a group of urban ethnographers who address a continuum of destruction to urban environments that includes natural catastrophes, urban terrorism, civil and pre-emptive wars in addition to traditional warfare on cities. They suggest "wounding" as an "organic metaphor" which "implies a vision of collective well-being that must be negotiated within an identifiable, bounded place" (Schneider and Susser, 1).

All the approaches outlined here pay particular attention to the interplay between global dynamics and local places in the contemporary world. In the following analyses, however, I step back in time to address the questions how films have historically addressed the politics of urban place and destruction in a primarily national framework; thus the films addressed here (except for those that are explicitly postmodern) situate the destroyed or divided city in its national context.

The city at war

Films about the effects of current wars in Iraq and Afghanistan, such as Bahman Ghobadi's *Turtles Can Fly* (2004), co-produced in Iran, France, and Iraq, and Mohsen Makhmalbaf's *Kandahar* (2001), co-produced in Iran and France, present desolate landscapes of disabled and traumatized refugees and survivors based on the destruction of cities that are not represented in the film. In the past, few cities have become a national reference point to rally around, either for the entry into war, as was the case with the US and Michael Curtiz's 1942

Casablanca, or for colonial liberation, as was the case with Gillo Pontecorvo's 1966 *Battle of Algiers*.

One of the most important films to mobilize a city is René Clément's *Is Paris Burning?* (1966). Set in Paris in 1944 during the German occupation, the film shows the fight of the Resistance against the Germans, who intend to destroy the city. It concludes with the ultimate victory of the Resistance liberating Paris. Because of the film's careful and detailed restaging of historical events, only the cast of international stars – Jean-Paul Belmondo, Alain Delon, Kirk Douglas, Glenn Ford, Gert Fröbe, Yves Montand, Anthony Perkins, Simone Signoret, and Orson Welles – betrays that the film is a historical re-enactment.

The city of Paris is the setting of the film as well as its subject and object. The French flag provides a national dimension; symbols of Paris such as the Eiffel Tower, the Seine, and Notre Dame appear in several shots; but the city is empty. Locations were carefully chosen, all signs of contemporaneity removed, and the urban landscape enhanced with signs and lampposts to look like 1944. This real setting validates the film's projected truth value and supports its national message of French triumph, but the film was created by an international collaboration, including a script by Gore Vidal and Francis Ford Coppola, and an international cast. Paris takes on a double function as the symbol for the French Resistance against German occupation, but in addition the emptiness of the urban landscape and the historical distance turns that landscape into a metaphysical site for a moral encounter in which self-determination and democracy win over dictatorship, violence, and destruction. The film was made at the height of the Cold War, four years after the Berlin Wall was built, when the moral impetus of the Second World War provided a rhetorical model of democratic (read capitalist) good against dictatorial (read communist) evil.

Is Paris Burning? aligns spectators with the Resistance through narrative and cinematic techniques. A contrasting example of a film that militarizes urban space can be found in Leni Riefenstahl's propaganda film *Triumph of the Will* (1935), which shows Nuremberg, a German medieval town chosen by the Nazis for their party rally, overcrowded with people celebrating Hitler. In the Nazi aesthetics of *Triumph of the Will*, the masses and the parades crowded into the small streets of Nuremberg are shot from above to create the majestic image of support for Hitler exploding the small, old-fashioned German town, so that the authentic setting of German tradition is modernized by the military apparatus and the arrival of Hitler in an airplane. *Is Paris Burning?* reverses this cinematic and ideological strategy by shooting from below, symbolizing the democratic Resistance against the dictatorship from above. The camera angle aligns the film's perspective with the underground movement and through this technological positioning aligns spectators with the Resistance.

Figure 5.1 René Clément. *Is Paris Burning?* (1966): The empty city shot from below

The major dynamic of this film concerns the mobilization of the people of Paris, and it increases as the film's narrative moves forward. The city's emptiness is gradually filled with masses rushing around, pasting posters proclaiming the liberation, taking over buildings, and riding bikes through the city delivering secret messages. The Resistance movement advances from the clandestine spaces of the cloisters and the underground sewer system into public buildings and from the night into the day. As the masses mobilize, the city comes alive.

The narrative of *Is Paris Burning?* functions in a national context but also in the transnational context of post-Second World War European cities. When the Resistance fighters discuss their options, one of them refers to the destruction of Warsaw, arguing that if they wait too long they will liberate only ruins. This perspective is juxtaposed to the Nazi occupation represented by General Dietrich von Choltitz, who is proud that it is the Führer's command to destroy the whole city. When he and the Swedish Consul discuss the fate of Paris, they frame the discussion in terms of "5,000 years of history going to the dust," negotiating occupation and Resistance in terms of the city and not the nation. The film's rhetoric of anti-fascism is based on a narrative of action against injustice, a

Figure 5.2 ***Is Paris Burning?*****: The city comes alive**

discourse of liberation and mobilization that is very different from the discourse associated with ruins which, as I will show in the rest of the chapter, is often associated with nostalgia and morality. The comparison with Warsaw as a city of ruins emphasizes the historical context of cities in Europe during and immediately after the Second World War, and also distances the political position of this film from the philosophical and metaphorical discourse associated with ruins.

The aesthetics of ruins

Ruins function in two distinct ways in films about war on cities: on the one hand, a setting of ruins claims realism, particularly in the rubble film; and, on the other hand, in postmodern film the trope of the ruin harks back to another genealogy from Romanticism and Baroque to contemporary film that does not entail realist specificity. Susanne Marshall traces different physiognomies of two different kinds of ruins: "The scars are inscribed differently into the ruin created by war and terror than in those old buildings created by the patient gnawing of the teeth of time" (46; my translations here and elsewhere unless otherwise noted). The latter, she suggests, is a threshold between culture and nature in the process of decay (44). Traditionally the ruin was a site of memory and contemplation. Marshall connects the pictures of ruins by painters Caspar David Friedrich (1774–1840), Hubert Robert (1733–1808), and Francesco Piranesi (1720–78) to the meditative films of Andrej Tarkowski, in which the characters slowly wander through ruins. According to Marshall the "aesthetics of ruins" shows that "concrete history demonstrates its power over the buildings created by human hand and again destroyed" (45).

Because the trope of the ruin relates back to a tradition of paintings, particularly in the Baroque style, where it functioned as allegory, Marshall argues that the ruin can also be read as marking the passing of time on a symbolic level instead of as a specific historical reference. So the trope of the ruin can function paradoxically to exceed the particular historical moment. Michael North shows how in English poetry written during the Second World War ruins take on a spiritual function when "the destruction of boundaries frees the imagination to roam an ambiguous territory between the historical and the mythological" (446). The immediate postwar rubble film employs the ruin primarily to claim a specific historical moment, but it can also move beyond the historical moment into metaphysical and moral meditations. The "nostalgic ruin" of the nineteenth century has turned into the "traumatic ruins" of the twentieth century (Assmann, Gomille, and Rippl 7).

The rubble film: the city in ruins

In the immediate postwar period in Germany, the screening of films was intensely public for two disparate reasons: Germans were forced to confront their collective guilt for the atrocities of the Holocaust in mandatory film screenings about the concentration camps; but they also went to the movies to escape reality and their crowded, destroyed, and cold apartments. The term "rubble film" denotes primarily films directed and produced in Germany directly after the Second World War, beginning in 1946 with Wolfgang Staudte's *The Murderers Are among Us* (1946) and ending with a satire of the rubble film, Robert A. Stemmle's *The Ballad of Berlin* in 1948. Robert R. Shandley lists 17 films for this brief period of two years (211–18). Most of the films are set in Berlin, including the three rubble films discussed here, Wolfgang Staudte's *The Murderers Are among Us* (1946), Gerhard Lamprecht's *Somewhere in Berlin* (1946) and Roberto Rossillini *Germany Year Zero* (1948). All three were produced by DEFA, which was founded in the Soviet Zone of occupied Germany in early 1946 by "a committee of Soviet officers, returning German expatriates, and resident German filmmakers [as] the first active postwar German film company," which subsequently became the only film company in the German Democratic Republic (GDR) (Shandley 17).

When Germany capitulated on May 8, 1945, the German film industry was destroyed, and during occupation the Allied forces created a licensing system for print, radio, film, and performance in their respective four zones of military government into which Germany and Berlin were divided. While the occupying force of the American sector, the Office of Military Government of the United States (OMGUS), emphasized re-education and economics, the Russian military government emphasized Germans re-educating themselves and ideological control. Despite the efforts of the Allied forces, some of the early DEFA films show aesthetic and technological continuities from the Nazi period because several directors and other filmworkers at DEFA had also worked in the Nazi film industry. DEFA films were supposed to present radical ideological and aesthetic breaks with the Nazi past, but were not always able to do so; and the same could be said of the West German film industry as well. The competition between the four Allied powers ultimately led to active licensing policies in all four sectors, since none of them wanted to lag behind the others.

Shandley defines rubble films as "a cycle of films" that address the problem of "the long shadow cast by the legacy of the Third Reich" and that "share the fundamental *mise en scène* of a destroyed and defeated Germany" (3). Many Germans had fled the cities and now returned to them only to find them completely different from what they had left behind. The Berlin of *Berlin: Symphony of a Great City*, which had consisted of entertainment, industry, traffic, and masses of people, had disappeared and given way to a landscape of rubble and ruins. Shandley points out that "[b]ecause of the special place film held as a propagandistic tool in the Nazi state and because of the economic potential Germany held as a market for Allied cultural products, the film industry was the last among the media to be allowed to re-enter the public sphere" (2). For Shandley, rubble film marks the period between the end of the war and the creation of the two German states (2), but he does not see it as an expression of "actual postwar German culture" because of "the omnipresence of the Allied censors in the filmmaking process" (7). Rubble film, he maintains, casts "certain versions of history into the public memory of that period, versions that then serve as models for how that past is remembered" (8). My concern here lies with the role that the city of Berlin plays in this construction of national memory.

Shandley approaches rubble cinema in the framework of German national cinema and thus addresses only films by German directors. However, it is striking that important rubble films were made by the Italian Roberto Rossellini (*Germany Year Zero*, set in Berlin) and by the British filmmaker Carol Reed (*The Third Man*,

1949), the latter set in Vienna. The city in ruins promises to capture the history of the immediate postwar moment by addressing audiences of Germans and/or Europeans; but, because the ruined city had lost its markers and has turned into an unrecognizable landscape, the rubble film also lends itself to negotiating abstract questions of morality, decay, and destruction, as is the case in *Germany Year Zero* and *The Third Man*.

Somewhere in Berlin, however, articulates the specific ideology of the Eastern part of Germany and captures the destruction especially of what was then the Eastern sector of Berlin. It begins with an abstract map of Berlin which, without any names, marks the city by the streetcar routes and the River Spree, reducing its built environment to its bare, one-dimensional outlines. The decentralized, natural shape shown on the map contrasts with the Nazi conception of the capital with a centralized plan to celebrate the fascist empire during the years prior to the film. Now everything looks the same and anyone could be anywhere, which levels the class differences between the pre-war Berlin neighborhoods, such as the expensive Westend and Kurfürstendamm, the communist "red" wedding, the Eastern Jewish neighborhood in Mitte, and the working-class neighborhoods of Kreuzberg and Neukölln in the West and Prenzlauer Berg in the East. The second shot then appears to be a well-composed establishing shot with Berlin's cathedral in the background, but when the camera pulls away this is revealed as an illusion, because the next shot shows men working, reconstructing buildings, and then looks down through a canyon of ruined houses and, in their midst, a makeshift market. Jaimey Fisher reads this shot as defamiliarizing central Berlin, an area presumably known not only to Berliners but Germans in general, which the shot "radically fragmented" so that the location lost its symbolic integration within the capital of Germany (469).

The film's title, *Somewhere in Berlin*, encapsulates the meaninglessness of local references for a city that has lost its markers, where people are displaced and social structures have fallen apart, housing is makeshift and claims to it are anarchic. Typical for the rubble film, the opening asserts historical accuracy before the narrative begins: we see rubble women forming a line and handing buckets full of rubble from person to person, and men are pushing carts instead of driving cars. At the same time, the landscape of the rubble serves as the site for important symbolic acts: chases that turn into fights between good and evil, and the death of children who symbolize innocence.

The characters in *Somewhere in Berlin* embody different relationships with the past. At the center of the film is the incomplete family (mother Grete and uncle Kalle) of the main character, Gustav, a young boy who is accompanied by his best friend, Willi. The two are part of a group of feral children who play war in the rubble and who eventually separate into two gangs. Willi has lost both his parents and is raised by Mrs Schelp, who owns a store and shares the apartment with Mr Birke, a black-market dealer who entices the children to buy his fireworks with food that they steal from their parents. Willi, in turn, steals Birke's stock of food to give it to Gustav as an anonymous gift for his father, who returned from the war hungry and hopeless. As a response Birke kicks Willi out and is arrested by the police for marketeering. Willi runs away from his makeshift home and is dared by other kids to climb up a ruin, from which he falls and subsequently dies of his injuries. As Gustav mourns the loss of his friend, his uncle Kalle explains to him that he needs a mission to help him get over his loss. Gustav gathers his friends together, and the final shot shows the children rebuilding Gustav's father's garage with picks and hammers when the father and uncle Kalle appear, illustrating the socialist ideology of working collectively to rebuild in order to overcome trauma. It is important in the narrative that the first building to be reconstructed is a garage and not a home: the place of production is privileged over the place of dwelling, reflecting the socialist primacy of labor. Put another way, in socialist films labor is the path to redemption. It is equally important for the ideology of the soon-to-be-constituted GDR that the children are the ones rebuilding the garage. They are the symbol of the future, and by the end of the film they have moved from immature reenactment of their parents' past – play-acting war – to the productive maturity of the new generation that will rebuild as a socialist collective. The title *Somewhere in Berlin* makes the claim that this kind of rebuilding could take place anywhere in the city that is to become the capital of the future GDR.

The connection between the ruins of civilization and feral children that underlies rubble films such as *Germany Year Zero* and *Somewhere in Berlin* continued in later DEFA films depicting teenagers in need of socialist socialization. They reflect the generation of children whose fathers did not return from the war, requiring the children to take on a mature role within the family. At the same time, this young generation can carry symbolic weight because it is not associated with the guilt of the Third Reich.

Berlin as the bombed former metropolis features in the majority of the rubble films, among them *The Murderers Are among Us*, which captures the immediate postwar moment in Berlin and by extension Germany. The film is one of the most pressing and self-critical in addressing guilt, responsibility, and revenge in the German postwar period, despite some of its problematic politics that I will address below. It begins with Dr Hans Mertens wandering the ruined city of Berlin on the way to a cabaret bar. The next shot shows us Susanne Wallner arriving on a train full of refugees and walking to her home through the city of rubble; she is returning from a concentration camp where she was a political prisoner. The two end up sharing Susanne's apartment. Mertens, a former doctor, cannot practice his profession anymore because he has been so traumatized by the war that he cannot look at blood, and thus – according to the logic of the film – he is a cynical drunk.

Susanne's neighbors in the building are a microcosm of postwar society, including Herr Mondschein, who has a little store and repairs glasses; Bartolomäus Timm, who reads the future for those postwar characters who are insecure and despairing about the present; a couple of gossipy female neighbors; and a couple of dancing girls from the cabaret. In a central scene Hans tells Susanne, as they walk through the rubble, that one day he will be able to say that he loves her but at that point he is unable to do so. The war has damaged his capacity for relationships and thus Susanne has to heal him. Susanne finds a letter addressed to Hans's Captain during the war, Ferdinand Brückner, to be delivered to his wife if he died on the battlefield. Susanne delivers the letter and finds out that Brückner is alive, but when she tells Hans, his memory of his traumatic past is triggered and we learn that Brückner was responsible for the killing of men, women, and children in Poland on Christmas Day, when Hans had asked him to spare them.

Brückner is exposed as a successful capitalist who profited from the war as much as from the immediate postwar period. Later Hans takes Brückner through the rubble intending to shoot him, but a desperate woman appears out of the ruins searching for a doctor. Hans performs an emergency operation on her child and so overcomes his own trauma by being productive. On Christmas he visits Brückner at his factory, apparently again with the intention of killing him. Hans confronts him while Susanne is at home reading his diary, which reveals the fateful events of the past to her and the viewer for the first time. Susanne subsequently prevents Hans's revenge on Brückner. The ending ties the love story and the meditation on guilt and revenge together in a final dialogue that proposes that it is our responsibility to judge and name the guilt of the past, but not to avenge it. Shandley reports that the end was forced by the Allied censors that were worried about vigilante justice – potentially rightfully so, given the outbreak of vigilante justice in other occupied areas such as Czechoslovakia.[11]

The film is framed by the invocation of death: it opens with crosses in the destroyed street and ends with mass graves. In the opening shot the camera moves out of the rubble, which reflects the psychological landscape, advocating the future of Germany and metaphorically leaving the rubble behind. Despite the accusation of guilt and the working through the past, the film rewrites German history dangerously without Jewish victims. The Holocaust appears in a shot of a newspaper headline at Brückner's house: "Millions of humans have been gassed." The film is thus part of a process of reducing Jewishness to a discourse, while trauma and suffering are presented via non-Jewish German characters, e.g. Susanne who is implicitly characterized as a former political camp inmate. The film confronts

Figure 5.3 Wolfgang Staudte. *The Murderers Are among Us* (1946): Walking through the landscape of ruins

the topic of guilt but embodies it solely in the figure of the bourgeois capitalist, while the figure of identification is the good German – a doctor – who has been manipulated in the war and is guilt-ridden. When the trauma is overcome, Hans's masculinity is reconstituted through productive labor. The turning point, however, occurs not in an institutional setting but in the midst of the ruins of Berlin. The ruined landscape is thus the site of destruction *and* healing, so that rubble films can, in the words of Fisher, "operate at precisely the intersection of personal memory and collective history" (472).

Hans's "invisible wounds" are in stark contrast to the hypervisible destruction in the city. Marshall contrasts the face of Hans, as "a human ruin in ruinous surroundings," to the bright face of Susanne and concludes: "The human physiognomy becomes part of the landscape of ruins; the realistic scene comments inevitably on the moral, spiritual, and physical constitution of postwar society" (46). The landscape of ruins becomes a symbol in the negotiation of morality with regard to reconstruction. Mertens says: "Rats, rats, everywhere rats, the city is coming alive again," a reference to Brückner and capitalist reconstruction. The image conjured up reverses the traditional understanding of ruins as associated with vermin and rubble, seeing them instead in the reconstruction of the city.

Germany Year Zero similarly situates moral decisions in the rubble and integrates documentary aspects into a feature film about the immediate postwar period in Germany, also beginning with a shot through the ruins, underlined by dramatic music. We see women and children working, an inspector picking out a boy who is too young to work, and children and grown-ups stealing coal that has fallen from a cart. The main characters are crammed into an apartment, and much of the narrative revolves around negotiating the shared space with each other, how to make do with limited resources, how to deal with the past, how to survive, and how to be a moral person amid corruption. The youngest son, Edmund, is overlooked by the grown-ups around him. His father is selfish and indulges in his own suffering, and his former teacher, Mr Henning, is coded as a National Socialist pedophile who manipulates Edmund to kill his father. At the film's conclusion Edmund becomes lost in the rubble of former monuments, meets abandoned, feral children, and returns to Mr Henning, to whom he confesses that he has killed his father, only for Mr Henning to slap his face and call him a monster. Edmund then kills himself by jumping to his death from a ruin.

The argument implicit in *The Murderers Are among Us* is diametrically opposed to that of *Germany Year Zero*, even though both films address the continuation of fascist thought. Whereas the former presents the continuation of fascism as capitalism, the latter Italian film approaches it as sexual perversion. In contrast to *Somewhere in Berlin*, in which the answer to the disorientation of the next generation is productive labor to reconstruct the city, in *Germany Year Zero* the legacy of fascism in sexual perversion leads to suicide. In all three films, *Somewhere in Berlin*, *The Murderers Are among Us*, and *Germany Year Zero*, the ruins and rubble mark historical specificity but also create a space for moral and metaphysical negotiation precisely because ruins enable abstraction from urban specificity.

The retro-rubble film

The rubble film is intimately tied to the postwar moment of the destruction left by the Second World War in the years 1946–48. In the early 1990s, Europe witnessed the emergence of films that I label retro-rubble, films which create intense nostalgia not for a specific and real moment, but for an imaginary, indeterminate past. The French film *Delicatessen* (1991) by Marc Caro and Jean-Pierre Jeunet disregards the historical discourse of post-Second World War but is simultaneously a product of that history. Not before 1990 could a film so closely invoke the traumatic history of the Second World War and at the same time disavow its specificity and historical weight by translating the traces of memory into absurd play. *Delicatessen* is part of the French *cinéma du look*, very much influenced by Hollywood and a departure from the French New Wave described in Chapter 3.

Delicatessen takes place in a ruined house in a post-apocalyptic but fantastic landscape populated by characters who represent absurd and exaggerated versions of Frenchness. A food shortage leads the inhabitants of the house to engage in cannibalism. An impoverished clown is supposed to be the next victim but he not only outsmarts the butcher who does the killing, but falls in love with his daughter. The ruin itself is obviously and fantastically staged as both dark and humorous, citing the bleak conventions of the rubble film and turning them into an endearing play of comedic action and ironic signifiers. In a reference to the conventions of the rubble film, children are witnesses in the setting of the house while the adults are primarily immoral. Members of the Resistance live underground in a fantasy space and use plungers to move along the wet walls, literalizing the term "underground." Their comedic representation goes to the heart of French national self-understanding and rewrites such important films as *The Third Man* (discussed in the case study) and *Is Paris Burning?* Only historical distance from the trauma of the Second World War and the disappearance of actual ruins makes the postmodern play of the signifier of the ruin possible.

Similarly, Lars van Trier's *Zentropa* (1991) returns to the German postwar period to create a slick, black-and-white, noir thriller that relies on recreating the mood and feeling of the postwar moment without being governed by the demands of the historical moment or geographical location that it invokes. The film employs the transportation system as its main site for addressing the past. *Zentropa* relies on a voice-over that narrates and addresses the American main character of the film and the audience, beginning as hypnosis: "My voice will help and guide you still deeper into Europa . . . open, relaxed . . . I shall now count from 1 to 10. By the count of 10 you will be in Europa . . . on the mental count of ten you will be in Europa . . . I say 10." During the opening voice-over the screen shows only train tracks, which in immediate postwar Germany signify transportation to the concentration camps. The narrative addresses the re-establishment of the train network in Germany and poses the question of how to reconstruct an infrastructure that is so representative of past horrors. The hypnotic rhythm of the film mirrors long journeys by train, back and forth. The film announces its historical and geographical place early on when the voice-over says, "From New York . . . you are in Germany . . . the year is 1945." That year becomes symbolic of defeat. The film creates a disorienting space, with hardly any recognizable cities. The film's plot line concerns the "Werewolf," a myth about Nazis who continued fighting after 1945, sabotaging the work of the Allied forces and liquidating Germans who cooperate with them. It addresses postwar anti-Semitism and shows Jewish returnees, but none of these characters is awarded any kind of interiority or subjectivity.

The film is not intended as realism. In several instances, characters crouch on the floor and a text is projected onto on the wall behind them, for example, when the word "WEREWOLF" appears in capital letters on the screen behind the main character. The film thus announces itself as an art film that references the specific historical moment and place but is not indebted to negotiating its precise politics.

Zentropa and *Delicatessen* stylize the periods they cite. In neither case is a city central, in contrast to the famous rubble and ruin films made immediately in the postwar moment. It is precisely the deterritorialization of the space that is evoked and its anonymity, either in the no-man's-land of *Delicatessen*, which refers to the city but never shows it, or in the train tracks as the permanent connection between different cities that are referenced but inhabit neither characteristics nor territory. The postmodern retro-rubble film relies on the abstraction of the city in which the rubble becomes a simulacrum of the immediate postwar moment invoking devastation without engaging with its politics or its trauma. *Delicatessen* needs to be not-Paris and *Zentropa* not-Berlin to emphasize the deterritorializing and detemporalizing aspect of postmodern, stylized ruins and rubble. Of course, though highly stylized, both films nevertheless speak to the politics of the early 1990s, at the end of the Cold War and almost two generations after the end of the Second World War, when the visible traces and the memories of that war's trauma were fading into the past.

The divided city

This section considers films set in divided cities, with a particular emphasis on Berlin because of its position as pawn and buffer between the former two superpowers during the Cold War. Divided Berlin created two different kinds of urban spaces despite its historical development as one city. Because Berlin occupied two states, the films discussed here were created by distinct film industries which created two distinctive urban cinematic aesthetics. Because the division of East and West Berlin was a process that took place throughout the postwar period, the filmic texts accompany the urban reconfiguration, provide ideological fodder for, and cinematically present ways to read the new urban environment to its respective citizens.

During the 1950s, films made by DEFA, the production company of the GDR, showed primarily young people torn between the seductiveness of the West and the socialist path of the East. These teenage characters continue the development of the children from the rubble film. These DEFA films stress the ideological differences between East and West marked by topographical and architectural differences. In both Gerhard Klein's *A Berlin Romance* (1956) and his *Berlin Corner Schönhauser* of the following year, West Berlin is identified with seductive movie-houses, unemployment, corruption, and sexual abandon. Both films end with the young characters following the path of socialism, which integrates "working, fighting, and loving . . ." as the voice-over describes the romantic couple at the end of *A Berlin Romance*.

The denunciation of West Berlin as a place of sexual and moral corruption because of American occupation is most pronounced in Karl Gass's *Look at this City* (1962), a DEFA propaganda film that justifies the building of the Wall as the "defense against imperialism." Over a shot of people streaming out of the subway station at the Hermannplatz (in Neukölln, in the West), the voice-over states: "People were lured over. Mainly young people. And saw this." The next shot shows movie theaters with posters for "skin flicks," and the following shots show comic-books. Film is associated with decadence coded as West German and lack of morals coded as American. Immorality is illustrated by a shot of a woman and her daughter in court with an inflammatory voice-over explaining: "They were not against West Berlin visits and comics." The voice-over accompanies a shot of comics with the word "theory" and a shot of a dead girl with the word "practice," and we are then shown images of a "murder weapon" and the "12-year-old culprit." West Berlin is thus associated with the lure of American popular culture that corrupts the next generation through sexualized visual culture. "Such political hypocrisy and depravity has to result in moral decay," says the voice-over with regard to images of a Parisian revue, female wrestlers, and dancing in clubs: "Once the symbol for moral decay was Chicago. Now it is West Berlin."

Thus the film justifies the presence of tanks in East Berlin on August 8, 1961: "Here it can go no further. This is the German Democratic Republic, where for the first time in German history peace is the program of the government." The voice-over proclaims: "Berlin is worth a war," an argument that *Is Paris Burning?* will appropriate for a democratic Paris four years later. In contrast to the decadent and corrupt space of the West, *Look at this City* illustrates "our socialist fatherland, the life and the future of our children," with images of East Berlin: children, women shopping, and students at the Humboldt University. The voice-over creates a defensive narrative of East Berlin being under attack, denounces West Berlin, but claims the whole city.

Kurt Maetzig's *The Story of a Young Couple* (1952) appears almost like a feature-film extension of *Look at this City* but was made a decade earlier. It makes the urban and architectural construction of socialist East Berlin its important visual and narrative turning point. Like *The Murderers Are among Us*, it begins with a Berlin in ruins. Agnes Seiler comes to Berlin through the snow in 1946 intent on starting a new life, but everything around her is drab. With increasing capitalist reconstruction, fascism resurfaces. Agnes, a positive, socialist–realist character, falls in love with Jochen and aligns herself with anti-fascist socialism. The film traces not only the characters' ideological development in the divided city, but the ideological debates concerning art that surrounds them.

The film's high point is the opening of the Stalinallee, in which Agnes participates. Urban development and socialist architecture are portrayed as "the cornerstone of a new, democratic Berlin" in the form of the Stalinallee, later renamed the Karl-Marx-Allee. During the opening of the grand avenue, Agnes recites on-stage a poem by socialist poet Kuba as part of the cultural program for the audience of workers:

> Peace has returned to our city in this street.
> The city was dust.
> We were dust and debris.
> We were tired to death.
> But how could we die?
> When Stalin has taken us by the hand,
> And told us to raise our heads proudly.
> We cleared rubble and made plans for green areas and blocks of houses.
> Then we were victorious and the city began to live.
> The path leading straight to Stalin is the path our friends chose.
> Never again will fire be reflected in these sparkling new windows.
> Tell me, how can we begin to thank Stalin?
> We gave this street his name.

We find in the poem and the cinematic staging a grandiose celebration of socialism, which extends to literature, architecture, and labor. The Stalinallee was the "German version of the Soviet paradigm of Socialist Realism," its "large-scale architectural forms" subordinated individual buildings to a grand vision of a "unified urban structure" with "functional" housing (Eisenschmidt and Mekinda, n.p.). The new socialist avenue symbolizes Stalin as father of the nation state which is imagined as part of a transnational communist community. The father figure of Stalin is doubled by another father figure in the narrative named Dulz, the father of Agnes's friend. He tells Agnes: "Now is the moment . . . good is surely going to triumph . . . All of us want the peace . . ." The family is mapped on to the political system, which expresses itself in the new urban topography.

My discussion of the cinematically divided Berlin has focused on examples made by the DEFA studio in which the representation of East Berlin coincides with the state ideology of GDR. Now let us look at films associated with Belfast and Beirut that were made by independent directors and that therefore foreground a biographical–political perspective criticizing the effects of political division on those who live in these cities. These two films, however, share with the DEFA films the centrality of the urban division for the narrative, the setting, the terrain, the built environment, and the psyche of characters that inhabit the city.

In *The Boxer* (1997), Jim Sheridan makes the division of Belfast productive for his argument about the politics of Nothern Ireland as well as central to the film's politics of remembrance. The film employs the topography of division for a radically different argument that the films driven by the state ideology of the GDR. The boxer Danny Flynn comes out of prison after fourteen years and is still in love with Maggie, who is now married to Danny's old friend, who in turn is in prison. Because Maggie's dad is the local IRA boss, the love relationship cannot be consummated. Danny was incarcerated because one of the most dogmatic and violent IRA members, Harry, abandoned him and Danny had to take the fall for him. Harry is killed at the very end of the film, punished for advocating senseless violence. Maggie's son Liam however is afraid that he might lose his mother to Danny and burns down Danny's gym, a place where Danny tries to bring Protestants and Catholics together. Because the dogmatic Harry is opposed to Danny's attempts at such reconciliation, he and his friends blow up a policeman's car during a boxing event that Danny puts on, and so end the hopes associated with the space of Danny's gym.

The melodramatic love relationship is negotiated via the *mise en scène* of the divided city. The film's love story takes place in different settings of imprisonment, beginning with the opening shot through the bars of Danny in jail, which is emblematic of the entire film. The film's opening connects prison to love and marriage on the one hand and to the militarization of civilian life on the other.

The opening, as with so many films concerning war, references the documentary convention by starting out with a statement of then president Bill Clinton followed by several statements of English and Irish citizens. Danny is a political prisoner – in the words of the IRA, a POW, a prisoner of war. This could be read as an indictment of state violence against Irish Catholics, but the narrative ultimately blames Harry, as the embodiment of dogmatic violence, for Danny's fate.

Belfast is portrayed through "grim concrete housing estates disfigured by graffiti, garbage, abandoned cars, and belligerent youths" (Cornell 83). Martin McLoone maintains that it is a "Belfast of the imagination," which does not rely on "realistic geography": the many shots from helicopters imply that "in the constricted space of the city the only freedom is upwards," and the manipulation of space reduces the city to a setting "of one block of flats in which everyone lives and fights and conspires and one square in which all the action takes place" (McLoone 2000: 77).

The Boxer manipulates the cinematic space in order to articulate the claustrophobic social relations between the characters; the general sense of confinement is expressed in shots from the helicopters of the wall dividing the city. The helicopters swirl above the action continuously, creating a sense of surveillance. Jim Sheridan intentionally left open who is in the helicopters, calling them agents, a kind of "big brother" (Sheridan, DVD commentary). The film's tight space centers on the gym in which Danny trains. All of it, inside and outside, is kept in a dark-blue tone which endows its space with nostalgia, which characterizes its impossible promise of reconciliation.

Only when Maggie and Danny go to east Belfast can the audience experience relief from the claustrophobic space that the film creates. Danny and Maggie sit on a bench surrounded by greenery. Yet the trip across the border is associated with danger and suspense. They are being watched and have to be escorted back. The beach outside the city is similarly associated with the burgeoning love of Maggie and Danny where they can talk about their love for one another. When Danny was in jail they could not have any contact and remembered each other's voices as confined in their heads. At the film's conclusion, after Harry gets shot, Maggie, Danny, and her son Liam tell a policeman at the internal border that they are going home. Thus, the happy end brings the relationship to a conclusion and also conjoins the two parts of the city in the notion of home, an imaginary construction that expresses the affective relationship to a place. The happy ending of their love relationship stands in for the hope that the city can overcome the political divisions.

Ziad Doueiri's *West Beyrouth* (Lebanon, 1998), like *The Boxer*, is intentionally political and revisits an important turning point in its country's history of internal division. While *The Boxer* emphasizes the ceasefire, *West Beyrouth* emphasizes the beginning of the civil war in 1975, when Beirut was partitioned along a Moslem–Christian line. Similar to the Berlin and Belfast films, the division also determines the narrative events in *West Beyrouth*. Like several of the Berlin films, *West Beyrouth* announces the division in its title and, like *The Boxer*, it begins with an invocation of documentary, but it also ends with documentary footage, integrating fictional and documentary material. While *The Boxer* is backwards looking to a time of hope with a nostalgic longing for the period before the escalation of violence, *West Beyrouth* returns to the moment of the inception of violence, in part, we can assume, to provide a personal history to contemporary violence.

West Beyrouth situates the coming-of-age story of three friends in the mid-1970s when they lose their innocence parallel to the city of Beirut. The children's generation has not internalized the division: Tarek is in love with the Christian girl May, and his friend Omar sticks by him. All three live in West Beirut. The young boys are shooting a film, which motivates their attempts to go to the other part of town to get their film developed. The three ride their bikes through the city trying to get to East Beirut where there is the only store that can develop their film. Soldiers try to hold them back and explain that only bombs go east, taxis go only to the border. Later Omar and Tarek join a demonstration without knowing what it is about. When the demonstrators are fired on, Tarek hides in a car and by accident gets driven into the eastern part of the city where he ends up in the mythical brothel of Oum Walid. When later he attempts to revisit the brothel with his friends we learn, with them, that the center of Beirut "is a no-go area." But Tarek has learned that in order to get to the brothel he has to walk through the area with the bra in the air, illustrating the absurdity of the situation. The owner of the brothel kicks the young teenagers out because one of the women had sex with two men of different religions, which destroys Tarek's hope for Beirut.

The film creates the city as militarized space increasingly divided, in which the exterior militarization and the interior tension mirror each other. Airplanes turn the sky into a militarized zone and the opening of *West Beyrouth* creates the school also as a militarized institutional space. The opening serves to construct the main character Tarek as a young troublemaker and situates the 1970s in the history of colonialization by the French, since he rebels against the French school system. France received Beirut with all of Lebanon after the collapse of the Ottoman Empire following the First World War, and Lebanon gained independence after the Second World War. When the civil war broke out in 1975, the city was divided into the western Muslim and the eastern Christian part, leaving the center an empty space (similar to Berlin).

The border interferes with the characters' daily lives and disorients them in the city. When the parents try to bring Tarek to school they cannot get through the policed border even though they invoke Beirut as their home town. They are told that "there is no Beirut, just East and West." Tarek's mother asks: "Which Beirut are we in?" and the father guesses: "West." This film foregrounds the characters' deep attachment to the city that goes beyond religious, political, or national affiliation. The exterior – the neighborhood – is slowly transformed into a landscape of war while the interior space – their apartment – is the site of the fighting between the parents. The film takes the convention of the coming-of-age story and complicates it through the political division of the city. At the end of the film Omar and Tarek reminisce how their lives have quickly changed from their innocent childhood, with which the film has started: Omar's sister ran away with a guy from East Beirut and his father went crazy, while Tarek's father is unemployed and hopeless and they have no money or food. Towards the end, they share mature and sad thoughts while bombs explode in the background. The film ends with a montage of vignettes of characters and documentary images of war, claiming a simultaneous realist and fantasy depiction of historical circumstances and individual attachments to each other and to their place of dwelling.

The cityscape of war articulates the dynamics of resistance against occupation; the rubble film becomes the setting for a moral engagement with reconstruction, in contrast to the retro-rubble film that invokes but disavows the past. The divided cinematic city is either used for a state-sanctioned position in state-produced films or as a biographical investigation of individual attachment and despair in relation to the divided city.

Case Study 5 **Carol Reed's *The Third Man* (1949)**

The Third Man brings together the two functions of ruins and rubble discussed above: to identify a specific postwar moment, and to create a landscape in which to negotiate moral questions. It was produced by Alexander Korda, who was born in Austro-Hungary but then became an important film producer in England, and the screenplay was written by Graham Greene; both men were active in the discourse around the Second World War in Britain.

The Third Man, based on a novel by Graham Greene and directed by Carol Reed, features Orson Welles, Joseph Cotton, Trevor Howard, Wilfred Hyde-White, and Bernard Lee, and was produced by David O. Selznick and Alexander Korda. It tells the story of Holly Martins, an American writer who arrives in Vienna to visit his friend Harry Lime but discovers that his friend is dead. At his funeral, Martins meets Harry Lime's friends and his former girlfriend Anna, and in the course of things hears differing accounts of Lime's death. Ultimately, he learns that his friend is not dead after all, and he arranges a meeting. In the meantime, the military commander informs him that Lime is involved in the black market, diluting and selling penicillin, and is therefore responsible for many deaths. Finally, Martins is persuaded to betray his friend, who is then trapped in the sewers of Vienna.

The film is famous for its noir aesthetics, but its particular narrative and moral appeal emerges from the connection between those aesthetics that make extended use of the underground sewer system and the city in ruins, as well as the immediate postwar setting of a divided and occupied city. In noir tradition *The Third Man* opens with a deep male voice-over in the simple past, connected to what appears to be documentary footage:

> I never knew the old Vienna before the war with its Strauss music, its glamour, and its easy charm. I really got to know it in the classic period of the black market. They could get anything, if people wanted it enough, and

continued

> had the money to pay. Of course, a situation like that does tempt amateurs but of course they don't last long, not really, not like professionals. Now the city is divided into four zones, you know, American, British, Russian, and French, but the center of the city, that's international, policed by an international patrol, one member of each of the four powers. Wonderful! You can imagine what hope they had, all of these strangers to the place, no two of them speaking the same language. But they were good fellows on the whole, did their best. Vienna doesn't look any worse than a lot of other European cities, bombed a little, of course. Anyway, I was dead broke when I got to Vienna. A close pal of mine had wired me, offering me a job doing publicity work for some kind of charity he was running. I'm a writer, name's Martins, Holly Martins. Anyway, down I came all the way to ole Vienna, happy as a log and without a dime.

The establishing shot shows us Vienna with its churches and classical statues as backdrop to the voice-over musings. But the footage of war-weary, haggard faces of old men exchanging clothes and wearing four watches on their wrists contrast to the splendor of times past. In a foreshadowing that can be understood only retroactively, we see a dead body floating in the Danube when the voice-over mentions amateurs involved in the black market.

Figure 5.4 Carol Reed. *The Third Man* (1949): The hunt for Harry Lime

The search for Harry Lime takes place in an urban landscape that is marked by the destructiveness of war, on the one hand and, the reality of a parallel, illegal economy, mapped onto the subterranean space of the city, on the other. As in other films noir, *The Third Man* is characterized by oblique camera angles and high-contrast shadows.

Figure 5.5 *The Third Man*: Skewed angles

When Holly Martins has to shout up to a building, he stands on a pile of rubble, and the façade of the bourgeois nineteenth-century house exposes its underlying brick work, gun-shot holes and damages from bombing. As in *Is Paris Burning?* the film's urban space is characterized by emptiness, which is historically accurate but also takes on an allegorical dimension. The shortage of gasoline and the overall destruction led to a lack of traffic in the cities of Europe. Thus, most of the cars we see in *The Third Man* belong to the military police. Ironically, however, cars are integral to the narrative and appear at important turning points.

continued

Figure 5.6 ***The Third Man*: Holly Martins on a pile of rubbish**

The setting of Austria, and of Vienna as its capital, becomes the backdrop for larger moral questions that flatten out the specific importance of Austria. Martins's voice-over that introduces us to the city reduces Vienna to its pre-war identity characterized by Strauss, glamour, and charm and to contemporary Vienna as "the classic period of the black market," which reflects the ironic distance of the film noir. "Vienna doesn't look any worse than a lot of other European cities, bombed a little, of course," explains his voice-over, which thereby erases the specificity of Vienna's historical and political situation. The shot of Martins standing on a pile of rubble calling up to a window could equally be set in Berlin or Paris. Yet, the political questions at that historical moment were distinctly different in Berlin, Paris, and Vienna. This generalization of Vienna allows the film to create a narrative that is both dependent on the postwar status of Vienna, as well as independent of its specificity in postwar Europe. The city in ruins represents the destroyed social structure of a city that makes unethical behavior possible. Yet resolving the issue of Harry Lime's black-market activities does not resolve Europe's, Austria's or

Vienna's future. In the final shot Martins waits for Anna, who walks away without speaking to him. Anna remains an enigma, her origin and her future unclear; she is a stand-in for the European refugee.[12]

The specificity of Vienna as Austrian is mostly disavowed, except through the music, which is played on a zither, traditional Austrian folk music. The confusion of the political situation is mirrored and expressed through the labyrinth underground. Bare brick, metal and water, and constant darkness characterize those scenes. The topographical labyrinth makes control impossible. Skewed shots emphasize that things are out of order. The world of the underground mirrors the psyche of Harry Lime in the noir aesthetics of ruins and the high contrast of light and shadow. The ruins and the underworld are associated with Harry Lime who takes advantage of the devastation; he thus functions narratively and symbolically as sign of danger, illegality, and threat to the healthy body of the city that the American military wants to guard.

The famous final shot of the pursuit of Harry shows a low camera close-up of Harry's grasping fingers through a grid of the sewer system. The majority of the shot is filled with the space of the empty city, with sparse light. The street, as

Figure 5.7 ***The Third Man*****: The return of expressionism**

continued

throughout the film, is emptied out of life. The city is haunted by Harry Lime's absent presence. His fingers, reaching out of the gutter, show the moment of his death. The shot harks back to the expressionist motif via film noir that has symbolically returned to Europe with the American Holly Martins. The grid of the gutter symbolizes the porous divide between the moral world above and the immoral realm of the underground, but the shot positions us above ground in the space of morality.

The Third Man employs the immediate postwar city for a noir suspense that also communicates a larger moral tale about corruption. That moral tale is played out in Vienna, but the film does not engage with the moral questions of the past for Austrians. The ruins become enigmatic, like Anna Schmidt, sharing noir's conflation of woman and the city, as explained in Chapter 2. The film's appeal lies in the postwar aesthetics of ruins employed for a claim to authenticity and a setting for a moral conflict.

Further reading

Stephen Graham (ed.) (2004b) *Cities, War, and Terrorism: Towards an Urban Geopolitics*, Oxford: Blackwell. This collection addresses new theoretical approaches to war and terrorism in relationship to a range of international cities.

William J.V. Neill and Hanns-Uwe Schwedler (eds) (2001) *Urban Planning and Cultural Inclusion: Lessons from Belfast and Berlin*, New York: Palgrave. These essays provide overviews and historical accounts of planning with an emphasis on cultural diversity in two cities characterized by urban division.

Jane Schneider and Ida Susser (eds) (2003) *Wounded Cities: Destruction and Reconstruction in a Globalized World*, Oxford: Berg. Essays that use the body metaphor to discuss the "wounding" of a range of international cities in the post-9/11 world.

Robert R. Shandley (2001) *Rubble Films: German Cinema in the Shadow of the Third Reich*, Philadelphia, PA: Temple University Press. The only single-authored book on rubble film in the context of German national cinema.

Essential viewing

René Clément. *Is Paris Burning?* (1966)

Ziad Doueiri. *West Beyrouth* (1998)

Carol Reed. *The Third Man* (1949)

Jim Sheridan. *The Boxer* (1997)

Wolfgang Staudte. *The Murderers Are among Us* (1946)

6 Utopia and dystopia: fantastic and virtual cities

> As we move farther and farther into the future, the probability is that the construction of new buildings will diminish, except in certain areas of the city, and the constant repairing, shoring up, and modernization of older architecture will begin to take on a rather retrofitted look.
>
> Ridley Scott

Learning objectives

- To be able to analyze the modernist vision of design and urban planning in early science fiction
- To account for the shifts of the portrait of the city in films that imagine the future throughout the twentieth century
- To be able to discuss the condition of postmodernity in relation to the postmodern city in contemporary science fiction
- To relate the changes of narratives in science fiction to the development of media technologies

Introduction

This chapter traces utopian and dystopian visions of urbanism in the cinematic constructions of fantastic and virtual cities, with an emphasis on the changing function of the city and futurity in science fiction, the genre that most directly addresses visions of the future. At the inception of the science fiction film, as for example in Fritz Lang's modernist *Metropolis* (1927), the city represented the future and was thus a prime site for the negotiation of utopian and dystopian visions. However, the late twentieth century saw two important shifts that changed this function of the city for the negotiation of utopia and dystopia. The city as a site of invention and innovation had defined labor as modern and industrialized in a capitalist system in contrast to rural, traditional, and premodern subsistence labor. In the late twentieth century, through the increase of computer technology, labor became independent of and detached from the city and turned into an invisible, deterritorialized, and solitary activity, while the medium of film began to incorporate advanced computer technology with computer-generated animation and computer game technology.

These recent developments in cinema – from analog to digital, from movie-going to home video, from film consumption to computer games, from celluloid to DVD – in conjunction with the changes in the nature of labor, have resulted in a detachment of the notions of progress and futuricity, whether utopian or dystopian, from the city. Now, even when their setting *is* urban, utopian and dystopian films focus on the conflict between human subjectivity and virtual reality. The city itself does not signify the future any more because futurity cannot be located in material technological development or in the built environment. Instead these contemporary, postmodern, science fictions narrate the difficulty of distinguishing reality and representation from one another.

To put it simply: the more we move into the future, the more these films show cities of the past or in decay. The chapter offers a historical outline, beginning with films from the 1920s that create modernist, futurist visions of the city and moving on to contemporary films that show decaying or outdated cities in the future that are often virtual: first, as the setting of the film that we are watching, but, second, within the narrative in the experience of the characters. This historical outline implies a moment in which films about the future show cities of the present, which I locate in the late 1960s to 1970s with the disillusionment of modernity, particularly for urban architecture, planning, and design, and with the onset of computer technology. This chapter emphasizes the analysis of films that represent turning points in this development from modern to postmodern cinematic depictions of the city.

Futuristic cities: *Metropolis* and *Things to Come*

Fritz Lang's modernist film *Metropolis* shows us a future in which utopian and dystopian possibilities are negotiated via the city. This Ur-text of the science-fiction genre was recycled with a Giorgio Moroder soundtrack (1984) and cited in Ridley Scott's *Blade Runner* (1982) and Tim Burton's *Batman* series (1989, 1992) (Brodnax 78). Shot throughout 1926, costing about 4.2 million Reichsmark and with over 35,000 extras, the film's narrative is set in the year 2000 (see Elsaesser 2000b: 29). The metropolis that gives the film its title is organized vertically: above the ground is the Garden of Eden for the sons and daughters of the owner of Metropolis, Frederson, and below the ground are the machines at which the workers toil, the workers' living quarters, and the catacombs where the disenfranchised meet. A few sites are not mapped on this vertical hierarchy: the house of Rotwang, the mad scientist, the cathedral, and the bar Yoshiwara.

In the fantastic and futuristic city of Metropolis, the workers toil below the ground without seeing the sun. Freder, the son of the master, Frederson, meets Maria who takes care of the workers' children, is enchanted by her, and follows her, first to a factory workplace where he switches places with one of the workers, and then into the catacombs where he hears Maria pray to the masses of workers about a peaceful revolution. Maria is trapped by Rotwang, the mad scientist, who creates a robot in her likeness. Frederson wants a robot to substitute for human laborers, but Rotwang intends the robot to incite a violent revolution and destroy Metropolis. The robot Maria is introduced to Metropolis in a seductive dance at Yoshiwara and then creates chaos among the workers, who begin to destroy the machines and leave their children behind in danger of being flooded. Maria and Freder rescue the children. The foreman recognizes the false Maria and the workers burn the robot at the stake. Freder fights with Rotwang on the top of the cathedral and Rotwang falls to his death. In the film's final scene Maria enables Freder to hold hands with his father, who symbolizes the head, and the foreman, who symbolizes the hand, thereby symbolizing the heart that connects the two.

The film integrates the discourse of religion, represented by the triangle of the three men in front of the cathedral, and of labor and revolution. *Metropolis* fetishizes the city and technology in a cinematic spectacle, emphasized in the opening shot of working industrial machines in abstract close-up shots. Shots of the city of Metropolis show a futuristic vision without humans but with airplanes and cars. The Bauhaus architecture at the time expressed modernity through practicality and rationality of design, but this utopian vision of progress is revealed in the narrative as a dystopia of panoptical control, disenfranchised and infantile workers, cruel exploitation, technological innovation gone mad, and self-inflicted destruction.

William Cameron Menzies's *Things to Come* (1936) similarly relies on set designs that create a modernist city which, according to Janet Staiger, is influenced by modernist architect and urban planner Le Corbusier's concept of the "Contemporary City." She describes the *Things to Come*'s "Everytown" of 2036 as "multi-storey buildings, moving sidewalks, and a domed glass shell protecting the climate, marking off city from exterior countryside and diffusing light to an even glow" (109–10). She suggests that *Metropolis* and *Things to Come* offer utopian visions that are "bright, optimistic views of possibilities for tomorrow" (111). Donald Albrecht describes how the city design of *Things to Come* was also intended to contrast with *Metropolis*'s urban future vision. Set designer Vincent Korda (brother of Alexander Korda, producer of *The Third Man*) remarked: "Things, structures in general will be great, yes, but they will not be monstrous" (163). Albrecht describes the futuristic city: "Korda's town center is beneath the ground and, with its smooth, plain contours, executed completely in white, is a conscious rejection of the frenetic pace associated with the Futurist-inspired cities of earlier films like *Metropolis* and *Just Imagine*" (164). Thus, *Things to Come* not only changes the set but also questions the implied hierarchy of the vertical organization in *Metropolis*. While the models of the two cities differ, both are situated in a "dialectical understanding of the relationship between the temporal and spatial dimensions of narrative utopia, a dialectic, that is at the heart of the experience of modernity as well" (Wegner xxi).

Because *Metropolis* and *Things to Come* project modernity on to the futuristic city, they also emphasize its material destruction. In *Metropolis*, the sequence in which the city itself is in danger of being destroyed is part of the spectacle of the film. Similarly, in *Things to Come* much of the narrative concerns war and its destruction of the city rather than the futuristic and utopian urban environment, which is quickly revealed as a dictatorship that leads to rebellion. Made in 1936, *Things to Come* is framed by its anticipation of war, beginning with discussions under the Christmas tree in 1940 about the relationship between war and progress in a city called "Everytown." The depiction of the war reflects the understanding of war in Britain at the time as a war without declaration. We see the shadows of soldiers marching across the screen, a city occupied by police and the military, trucks full of people wearing gas-masks driving around, and seemingly random explosions. The depiction very much reflects the anxiety about a war in England, particularly as it pertained to London. The city is, however, not London but one that looks like London. The extended scenes show total war in montages of explosions, airplanes, people running in panic, and close-ups of the faces of a dead child in the tradition of the anti-war film.

In contrast to *Metropolis*, *Things to Come* ties technological progress to war and not to capitalist production. War centrally organizes *Things to Come* and thus it not only takes up extensive screen time but also extensive narrative time, moving from 1945 to 1960, when a National Bulletin announces "Victory is Coming." In 1966, Everytown is destroyed, and we see the ruins in shots that are reminiscent of the films analyzed in Chapter 5, ruins which will be rebuilt as the utopian city in 2036, described earlier. Each film offers a visual spectacle of a modernist city that captures a utopian future in its setting, while the narrative engages critically with dystopian visions of a modernist future.

The totalitarian computer

With the advancement of computer technology the importance of the cityscape in science fiction decreased because visions of the future became associated more with the invisible aspects of the technology and less with visible design and urban structures. In Jean-Luc Godard's *Alphaville* (1965) the computer controls the city of that name. *Alphaville* is a transitional film that represents a shift towards virtual reality, while it is beholden to a realist representation of the city. Therefore, for example, the computer takes up enormous space in the heart of the city. *Alphaville* signals another transition related to the paradoxical development of the postmodern city becoming a city of the past, namely the genre mixing of an ostensibly science-fiction film shot in a noir style with a main character, a "private-eye" taken from American pulp fiction of the 1930s and 1940s. The film mixes science fiction and film noir conventions to criticize urbanization as inhuman. Much of the main character Lemmy Caution's time in Alphaville is spent in a hotel room, a transitional space typical of film noir. Kaja Silverman, in dialogue with Harun Farocki, describes the dystopian element of Alphaville as a "technocratic vision" (62), but the setting of the room and the "Red Star Hotel," according to Silverman, "is a throwback to a detective novel from the 1930s, or a B-picture from the 1940s" (66).

The FBI agent Lemmy Caution comes on a mission to a futuristic city and takes on the identity of journalist Ivan Johnson working for *Figaro–Pravda*. Alphaville is run by brainwashed automatons and Alpha 60, a large computer. The city we are shown looks just like Paris, but without the typical signifiers of Paris, "in the process of radical modernization" (Forbes 50). Lemmy Caution fights Alpha 60, a totalitarian system of rationality, and awakens Natasha, one of the automatons, to her feelings and then escapes Alphaville with her. Alphaville had been constructed by Leonard Nosferatu under the name of Leonard von Braun, a man of science, who now controls the city by commanding the computer he has created. Alpha 60 represents the danger of destroying Alphaville, and only Lemmy Caution can outdo the machine.

The city that signifies the future is recognizable as Paris of the 1960s. The alienation effect here is not the estrangement of a city through design that envisions something radically new but instead lies in the film's suggestion that we see the familiar city in unfamiliar ways through a narrative that is projected on to the cityscape. Similar to the changes in perception of time and space in modernity, these categories underwent seismic shifts in postmodernism, and *Alphaville* stages, I argue, an encounter of modernity and postmodernity. Silverman explains the relationship between computers and temporality in *Alphaville*: "To be a computer is to be simultaneously outside of temporality, and made up of memory. It is to be incapable of retrospection, and yet unable to forget" (70). *Alphaville* introduces the conflict between affective human and technological memory that continues through the later films discussed in this chapter.

The film follows some of the conventions of the science-fiction genre but estranges them through its settings in everyday Paris, the use of real names (von Braun) and names that reference film and literary history (Nosferatu). The combination of these references projects a postmodern pastiche onto a traditionally modern urban setting, which creates a sense of disjointedness between setting and narrative. The narrative parts are paradigmatic pieces from the genres of the private detective and science-fiction films. Thus, while the film's setting is a coherent and whole city, the narrative conventions, as well as the names, create narrative pastiche.

Alphaville is not another planet; according to Silverman it is "a state of mind, a 'place' where people find themselves when reason has succeeded in driving out affect" (77). While she is correct in her assessment that Alphaville represents a state of consciousness in which rationality has conquered emotion, it is important to note that this state of consciousness is still represented by a city, and a well-known and recognizable city, for that matter, Paris. The film gives the city of Paris, which we see and know and is inscribed by the French New Wave as authentic, the name of Alphaville, the first city. Thus, the film does not offer us an abstract meditation on temporality and consciousness, but instead anchors these two issues in a recognizable cityscape. This move continues the tradition of the science-fiction genre that situates the conflict between the poles of extreme rationality and spiritual emotionality in an urban environment. The films discussed in this chapter do not simply suggest cities as sites for this conflict but instead propose cities both as origins of the conflict and as necessary sites for negotiating the contradictions and tensions that emerge from the conflict. Curiously, despite Godard's *avantgarde* status, the ending follows conventional narrative structure, discussed by Silverman who concludes about the space of the city circumscribed by the film *Alphaville*:

> At the end of *Alphaville*, Lemmy and Natasha travel through the night in the Ford Galaxy. Significantly, they never reach the geographical border where Alphaville ends, and our world begins. That is because the border is psychic rather than terrestrial – because the earth is a state of mind, rather than a place.
>
> (82)

I want to suggest here *Alphaville* reflects the moment when notions of temporality and space underwent a change from modernity to postmodernity, similar to the changes to space and time in modernity, described in the Introduction. Later films discussed in this chapter, so-called "edge of construct films," leave spatial coherence and temporal continuity entirely behind through the use of virtual reality. They create a viewing experience in which temporality and spatial markers are disorienting; *Alphaville*, in contrast, suggests problems of space and temporality in its narrative but presents coherent images of a whole city and a linear narrative set in an undefined time period.

The superhero in an outdated city

Richard Donner's *Superman* (1978) connects futuristic aspects of science fiction with the setting of an outdated city and begins the paradoxical portrayal of cities in futuristic narratives in which they appear to belong to a contemporary or earlier time period. In this film, the city becomes the dystopian reflection of anxieties about urban development at the time. The futuristic and fantastic utopian aspects associated with science fiction are embodied by a superhuman individual who is, not coincidentally, male and whose masculine prowess is put to the service of the nation to which the city is subordinated.

In early science fiction, such as *Metropolis* and *Things to Come*, the design of the *mise-en-scène* of the city represents the future where utopian and dystopian visions of the social world are negotiated. *Metropolis* and *Things to Come* articulate a political dystopia to critique certain aspects of modernity and they celebrate modernism's utopian, technological spectacle in the design of the imaginary city. In *Superman* the city itself is not part of the futuristic setting; instead the character is endowed with superhuman strength to enable utopian moments that are integrated into an overtly nationalist discourse. The city is situated in the present in a historical moment when the actual cityscape lost its importance for science fiction because the built environment became less associated with modernist, utopian (and thus also dystopian) visions of urbanism and instead became the object of technological surveillance. *Superman* ideologically supports surveillance in the figure of the all-knowing, all-seeing character who works for the good of the nation.

Superman integrates the binaries that surround the context of the city: the urban–rural split and the future–present binary, both dialectically embodied in the dual nature of the nerd–superhuman. *Superman*, which takes place primarily in New York City, begins with Superman's biographical pre-history, the background that explains Superman's superhuman strength, situated in a different universe, on the planet Krypton. Here we find a futuristic, technological society with utopian elements, such as strength, knowledge, technological, and architectural advancement in the representation of the city and dystopian elements, since the politics of the society are cruel and lead to its destruction. The parents of the baby, who will become Clark Kent, alias Superman, send him into space while Krypton explodes. They provide him with an entire knowledge of the galaxies, and in voice-over the father explains the child's superpower on earth.

The futuristic city on Krypton, many galaxies away, is contrasted to the rural environment on earth where future Superman lands. An elderly couple finds the baby, takes him home, and raises him as Clark Kent, instilling morality, humility, and a sense of undifferentiated religiosity in him, all thus marked as positive midwestern American values. The interior of their home, as well as the barn and the field, all signify traditional, middle-American values juxtaposed to the city. In a fantasy of assimilation, *Superman* creates the figure of Clark Kent who passes as a nerd but has a secret alien identity endowed with superhuman strength. He moves from the American midwest into the city, the destination of migration. Before Clark Kent enters the city, however, he connects with his past in the extreme environment of the North Pole where he receives his Superman robes, with which he can fly. He comes to the city as a shy, nerdish young man, and gets a job at the *Daily Planet* newspaper where he falls in love with Lois Lane, a self-confident, feminist, urbanite journalist. He is able to seduce Lois but only in his incarnation as Superman when he represents the asexual, yet perfect masculinity that espouses old-fashioned values of gentlemanly and well-mannered behavior. As an urban intellectual, he is invisible to her. *Superman* denounces qualities associated with urbanity, including social movements, such as feminism, by implying that Lois's real desire is for protection and romance.

Superman's persona is explained by the combination of his Kryptonic superhuman strength and the rural, midwestern upbringing that gives him moral integrity. The film contrasts two cities, one on planet Krypton that is technologically advanced and the other on Earth that appears to be New York City in the 1970s that consists of media, multicultural everyday life, feminists, and a criminal underworld that has access to technology. Clark Kent's upbringing in wholesome, rural, middle America is sandwiched between those two urban places, one in which technology is coded as positive and one in which it is associated with crime. The film portrays rural, middle America nostalgically, a paradox in a film that relies so heavily on digital effects for its science-fiction narrative.

The nostalgia that shapes the representation of the American midwest also characterizes the film's opening of an old-fashioned movie-house screen including curtains. On the screen within the screen, we see a comic magazine and a hand that turns black-and-white pages. A child's voice-over situates the narrative in the 1930s, but addresses those who do not know the cultural, political, and historical signification of the 1930s Depression:

> In the decade of the 1930s even the great city of Metropolis was not spared the ravages of the world-wide depression. In the times of fear and confusion the job of informing the public was the responsibility of the *Daily Planet*, the great metropolitan newspaper whose reputation for clarity and truth had become a symbol of hope for the city of Metropolis.

The voice-over invokes a historical framework but also references comics, the origin of the Superman story, which is updated in the *Superman* films. From here, the film moves into the drawn comic on the page and from the drawn images into space and to Krypton. The futuristic environment of Krypton is reminiscent of the convention of the futuristic city as portrayed in *Metropolis*. The white setting and the crystals signal purity, which will be repeated as a motif in the North Pole on Earth where Superman's well-known robes originate. But once the narrative of Clark Kent begins and he comes to the city, the time period evoked by robes and *mise-en-scène* is not of the 1930s but the 1970s, and the city evoked is not some futuristic metropolis but contemporary New York City. Thus, *Superman* aligns comics and the 1930s with futuristic cities and film with contemporaneity that no longer associates cities with a utopian future. The opening implies that in an earlier period that connoted modernism, utopian thinking could envision a city of the future, but that these utopian visions have given way to a pragmatic realism and that only an extraterrestrial, superhuman power can protect cities and the nation from crime that originates in the city.

Superman continues the motif of the vertically divided city introduced in *Metropolis*. Here, the city is vertically divided into a space above and below, like *Metropolis*, but the space below is associated with criminality carried out by Mr Luthor, Ms Teschmacher, and Otis. The criminal life underground is portrayed as excessive, endowed with bourgeois, nineteenth-century, decadent kitsch, with artificial light, Hawaiian music, and books, in contrast to the newspaper associated with good, modern people. This reverses the values associated with the vertical organization of *Metropolis* where decadence and bourgeois trappings are associated with the upper world of the ruling class. Whereas *Metropolis* critiqued the ruthless exploitation of the working class by capitalists, *Superman* projects the accoutrements of the bourgeoisie onto the criminal elements of city.

The showdown between Lex Luthor and Superman centers on the increase of real-estate value in California, which brings panic to suburbia. The film creates a set of spaces coded as integral to the American social landscape and fabric, the wholesome heartland, the decaying city, the decadent underworld, and dangerous nature in the desert. Superman delivers Lex Luthor and Otis to jail; when the warden thanks him, he answers: "Don't thank me, Warden. We're all part of the same team." His statement, "I'm here to fight for truth and justice and the American way," clarifies that his presence is necessary because the city and its urban inhabitants have lost the American way.

Despite the film's opening's announcement about the depression in Metropolis, the city in *Superman* is contemporaneous with the time the film was made, and signifies decay and criminality, in short, the failure of modernity. In 1972, the urban housing project called Pruitt-Igoe had been blown up in St. Louis, Missouri, which postmodern architect Charles Jencks saw as the symbol of the end of modernism. Pruitt-Igoe was a modernist complex that included over 2,000 public housing units that had been built in 1951 by architect Minory Yamasaki. By the late 1960s–early 1970s, it was clear that these urban projects of modernist visions had failed and the only response the city found was total destruction. Positioned after what was perceived as the failure of modernist architecture and urban planning to embody a utopian vision of the future, *Superman* espouses conservative values by subordinating the city to the nation, portraying the battle between good and evil, and portraying an asexual masculinity that romances and seduces the urban feminist.

The virtual city of the past

Whereas the early films of modernist science fiction portrayed dystopian and utopian visions in futuristic city settings, and films from the 1960s and 1970s show contemporary cities marked by dystopian narratives, the late 1990s saw a group of films that showed cities of the future – either real or simulated – that were either decaying or belonged to the past. I read these films and their cities as marking the passing of the promise of modernism and diagnosing the condition of postmodernity. Joshua Clover identifies the collapse of virtual and past cities in what he labels "edge of the construct films" (see also Sobchack 1999): films in which "the hero sees the simulation as nothing more (and nothing less) than what it is, recognizes the limited apparatus of what he once thought was infinite reality" (8). He describes the cities in three films from the late 1990s as "constructs in the past" even though they are virtual cities: Truman's town in *The Truman Show* (1998), which is set in the 1950s, the "pointedly motley but largely 40s noir metropolis" of *Dark City* (1998), and the "1937 Los Angeles" recreated by programmers in Josef Rusnak's *The Thirteenth Floor* (1999) (9). While the city was central to the invention of the science-fiction genre and to the negotiation of utopian and dystopian visions of the future in Lang's *Metropolis*, the significance of the city as a recognizable, even if fantastic, localized, built environment to the negotiation of utopia and dystopia diminishes with the advancement of technology. Instead, recent films are set in urban environments that are decaying or are from a by-gone era.

The contemporary dystopian vision concerns the problems of reality, virtuality, memory, and subjectivity. Paradoxically, while cinematic representation is enabled by the technology, it is technology itself that becomes the problem in negotiating questions of utopian and dystopian visions of the future. Andy and Larry Wachowski's *The Matrix* (1999) shows that the science-fiction genre has undergone significant changes as technology has developed, allowing other genres to employ similar technologies so that genre boundaries have become less distinct. Of course, science fiction has never been absolutely pure; a film like *Metropolis* is also a love story and a melodrama with religious overtones that addresses, through a futuristic narrative, the conflict between labor and those who own the means of production. *The Matrix*, in turn, integrates Hong Kong action films, a genre to which the physical body is integral (see Chapter 4), with technology associated with video games.

The oppressive quality of technology, which threatens to empty humankind of its humanity and is represented by robots, replicants, and automatons, then resurfaces later not only in *The Matrix*, but also, for example, in Ridley Scott's *Blade Runner* (1982), and in Alexis Proyas's *Dark City* (1998). Paradoxically, these films fetishize, celebrate, and rely on their own technological advances in camera and animation technology and in special effects. In these later films, however, the built environment of the city is no longer the site of modernity and technological innovation, but a grimy place of the present and the past that has more in common with the city of film noir (see Chapter 3), and therefore sometimes called "future noir" (see Sammon; Staiger), the city in ruins (see Chapter 5), or the rundown urban ghettos and barrios (see Chapter 7). Portraying an outdated or dilapidated city is part of the general dystopian vision of technology. As technology advances in society, we increasingly find artificial humans, cities, and spaces in the urban science fiction. Science-fiction film is caught in a curious paradox: the more the advancement of technology lends itself to narratives fed by anti-technological anxiety and conspiracy, the more the representational strategy can rely on technological development. The cities in *Dark City*, *The Matrix*, and *Blade Runner* (see Case Study 6, pp. 45–48) are dystopian sites of decay based on seeing technological advancement not as utopian fantasy but as extreme dystopian fantasy.

Dark City includes explicit references to the history of modernity. It references modernity in the design and architecture of the cityscape, but also in the interior design and clothing, and via one of the main character's names: Dr. Daniel Paul Schreber. Daniel Paul Schreber was Freud's patient in his famous case about paranoia, about which Schreber himself wrote a book (on the Schreber case and modernity, see Santner). The reference to turn-of-the-century Vienna and the beginning of theories on paranoia is ironic because the film itself offers a paranoid fantasy.

Like so many of science-fiction films, *Dark City* begins with a pre-history of a different civilization. The voice-over informs us: first there was darkness, then cubing. Their civilization was in decline. Then the camera moves down into the city echoing the beginning of *Metropolis*. The city of *Dark City* is futuristic, like *Metropolis*, with artificial skyscrapers, but it is also outdated in terms of architecture and decorative style, which are reminiscent of the 1940s and 1950s. Except for Dr Schreber everything stops at midnight, referencing older narrative structures, particularly turn-of-the-century fiction and fairy tales.

In the story of *Dark City*, a glitch in the system has the effect that Mr Murdoch does not fall asleep at midnight and now knows that every night the city is reconfigured and its inhabitants injected with new memories. His wife searches for him following his disappearance, but since she may have been given a false memory, it is not clear that she is really his wife. Six prostitutes have been killed, and the cases are the responsibility of Inspector Bumstead and Walenski, a former police officer who went mad. Because Murdoch was not given a new memory at the last cubing, he develops the power of the aliens and can use it against them. *Dark City*, *Blade Runner*, *Alphaville*, and *Superman* are all concerned with memory as our capacity for maintaining subjectivity, but it is precisely in its capacity for memory that the computer has the advantage over humans. All three films question affective memory as a characteristic of humankind: how does the way we remember differ from the computer's operation of memory and what would happen if that difference were to be lost?

In *Blade Runner* the replicants remember a made-up family history via staged photos that give them access to stored emotional responses. In *The Matrix* the main characters seem to live a life without affect. In *Dark City* the trigger for Murdoch's memory of Shell Beach is a postcard that leads him to a poster on a brick wall – behind which is nothing but the empty universe. Typical of the postmodern condition, the trigger of an intimate memory of one's past is found in advertising, a simulacrum of intimacy, which he searches for the remainder of the film. *Blade Runner* and *Dark City* use photos and postcards as triggers for memory and in both cases those visual representations, which lead into the past in *Blade Runner* and outside of the city in *Dark City*, are false. These forms of visual representation self-reflexively comment on the medium of film and its capabilities.

This significance for postmodern representation can be highlighted in a comparison to Truffaut's self-reflexive use of himself in the carnival ride that looks like a zoetrope in *The 400 Blows* (discussed in Chapter 3). Whereas Truffaut's presence endorses the film's recreation of the affective aspect of memory by documenting the authentic city, *Blade Runner*'s and *Dark City*'s references to the visual medium suggests that visual representation has become unreliable.

The fact that the city, the built and lived environment, changes everyday mirrors the lack of memory. In an allegorical reading, the film asks how our human memories can be maintained in the face of the onslaught of urban change and virtual reality. In contrast to *The Matrix* and *Blade Runner*, however, *Dark City* posits the human body as the container for truth because it is real. Jonny Murdoch visits his uncle and learns that his parents died in a fire and that he has a scar from that fire. His body, however, has no trace of a scar. The absence of the scar proves that his memory is false; however, in a postmodern move, he embraces the false memory because it is the only memory he has.

However, Murdoch wins against the aliens, the sun comes up, and he meets a woman with whom he walks to Shell Beach. The film thus beautifully illustrates contemporary postmodernism as outlined in the Introduction. By self-reflexively questioning the truth-claim of visual representation, the film endorses not only the paranoia of the characters but also of the viewers. By embracing the simulacrum of Shell Beach, the film proposes the absence or insignificance of any real locations. And, finally, by having an advertisement for a simulacrum, the film shows postmodernism's ties to late capitalism.

Dark City connects 1940s noir and science fiction in its setting, which repeats the genre-mixing of *Alphaville*. The designers gave the city an "anachronistic feeling," creating a script filled with "period Americanism" (DVD commentary). The film combines live action with a planet created digitally from scratch. The more technology advances, the more the cinematic city is imbued with nostalgia, even if that nostalgia is created by advanced technology. The question of the real and the simulation is, however, not only a philosophical question related to postmodernism but an economic question connected to the increased circulation of computer games that are tied to these feature film releases. So, in some ways the feature film functions like the postcard that sends the main hero on a search that can be continued infinitely in the virtual world of a computer game. If we imagine the spectators of these films also as consumers of the accompanying computer games, we can also understand that the city and its reality are not important any more. The ideal spectator is imagined not as a consumer of a film in a movie theater in the entertainment district of a modern city, as in *Berlin: Symphony of a Great City*, but instead as a solitary consumer of videos, DVDs, and computer games. The ideal spectator is inscribed in the film as a male to whom the question of whether his wife is really his wife or whether a woman at the beach is really his old love is no longer of significance. Instead, for ideal consumers of virtual realities, the exchangeability of figures and places reflects the pleasure of the game within the virtual world. This is also the case, for example, with regard to the next film discussed here, *The Matrix*.

Caught in the matrix

By the time of *The Matrix*, the built environment of the city is the location for neither utopian nor dystopian visions. The film's urban scenes are shot in Sydney, a neutral city in the global imagination, since it carries no connotation of utopian urban development or urban blight. With the advancements in digital media, hyper-environment, and virtual reality, the city is no longer the embodiment of futurism. Joshua Clover convincingly argues that the reorganization of labor in the late 1990s, particularly in the US, found expression in the paranoid logic of *The Matrix* (see 71–4). Importantly, labor has become detached from locations and can take place anywhere, thus not necessarily in the city.

The city we encounter at the opening of *The Matrix* shows dilapidated blight marked as the present. The apartment of the main character, Neo, is dingy and could be in a tenement house in New York City. Similarly, the spaceship *Nebuchadnezzar* is outdated, in curious juxtaposition to a narrative about a computer that has total control. When Morpheus explains the matrix to Neo, they are meeting in a grimy, old apartment, with Morpheus sitting in a dilapidated chair. Morpheus's explanation of the matrix encapsulates the dystopian vision of technology:

> The matrix is everywhere . . . it is all around us . . . you can feel it . . . when you to work . . . when you go to church . . . it is the wool that is pulled over your eyes . . . you were a slave . . . born into a prison that you cannot smell or touch . . . a prison for your mind . . . unfortunately no one can be told what the matrix is . . . you have to see it for yourself. . . .

The condition of slavery has changed from that of the workers underground in the city in *Metropolis* to that of people who are in a virtual prison of their own minds.

Neo follows Morpheus to his spaceship *Nebuchadnezzar* after he is given the choice of a blue pill or a red pill, one of which would return him to his prior innocent sense of reality and the other which would lead him on the harder path of wisdom. After he is tied to a chair and advanced into another state, Morpheus explains that they are now inside a computer program and that what we see is a mental projection of the digital self. "What is real? How do you define the real?" he asks. Then he shows Neo a computer screen and tells him: "This is the world as it exists today, the desert of the real." He shows Neo, and thus the audience, the dystopian vision of science fiction: the urban space of ruins on an outdated television screen in black and white. In the twenty-first century artificial intelligence cannot survive without an energy source (as in *Dark City*), so the machines have found a way to get the energy they need by growing humans instead of humans being born. The matrix is a computer-generated dream-world built to keep humans under control. When Neo is in the matrix, he can learn martial arts. When he has to visit an oracle, he finds her in a dilapidated urban housing complex. *The Matrix* is a postmodern combination of genres, such as science fiction and Hong Kong action martial arts film. In the final shoot-out, we see the famous bullet time – a computer-animated martial arts move – in the subway station. The genre's action takes place in an urban environment, but the setting of the city in the future looks like the city of the present in a process of decay. While modernity was expressed in modern design, architecture, and urbanism within the film, postmodernity is expressed in a pastiche of genres and settings. The conflict concerns the question of representation for the characters but also for the audience.

Almost saved by the mall: dystopian horror in suburbia

Not belonging to the genre of the science fiction but rather to the horror genre, George A. Romero's *Dawn of the Dead* (1979), remade in 2004 by Jack Snyder, creates a fantastic dystopian vision of the new American urban development, suburbia. Zombies invade suburbia and force the film's main characters to find refuge in a shopping mall. Here the most normal space of consumption, intended to calm and seduce citizens into consumerism, becomes the alienated site of horror, where the main characters are threatened by cannibalistic death. All those mall characteristics, such as elevator music and artificial fountains, are estranged by the horror effect. The characters begin to live artificial lives in coffee shops, furniture and electricity stores watching television, imitating the lives of consumption in the 1970s. *Dawn of the Dead* turns the shopping mall – the promise of a safe and clean haven of consumerism – into the horrific site of cannibalism. The zombies are the suburbanites gone bad, reverting to the base instinct of killing through biting. The subversive impetus of the dystopia lies in its reference to the everyday of modern consumer society, in contrast to *The Matrix* that ties its narrative to religion, in a way similar to *Metropolis*, which promises redemption. Postmodern cities in science fiction are in the process of becoming outdated, while the contemporary urban development of malls is subversively portrayed with ironic horror as a site of danger, a counterpoint to the *Superman* national ideology of the late 1970s, and the endorsement of virtual reality in the late 1990s.

Case Study 6 **Ridley Scott's *Blade Runner* (1982)**

Ridley Scott's *Blade Runner* (1982) is a postmodern science fiction film that explicitly references the modernist classic *Metropolis* in its set design, narrative, and themes. Rick Deckard lives in the retrofitted Los Angeles of 2019. As Blade Runner, he hunts replicants and administers the Voight–Kampff test to identify them. The film creates doubt about the status of replicants and humans, including the main character Deckard and his love interest, Rachael. When six replicants have escaped from off-world and are now roaming the streets of Los Angeles, Deckard hunts them in order to kill them. The replicants have come to Los Angeles to visit the Tyrell Corporation, where they were made, because they want to live longer.

Blade Runner continues and updates the issues *Metropolis* introduced. Like *Metropolis*, the film opens with a text that scrolls across the screen and that explains that replicants work off world in slave labor. The *mise-en-scène* of *Blade Runner* integrates futuristic city design with flying vehicles, such as the police spinner, neon advertisements, and billboards. And, like its precursor, *Blade Runner* celebrates the design of the city, while the narrative portrays a dystopian vision of humans being replaced by replicants. Similar to *Metropolis*, technology created by humans has erased the recognizable difference between machines and humans. Replicants have memories stored that they can call up with photos.

Blade Runner shares with *Metropolis* the importance of the *mise-en-scène*. The city is postmodern but also decaying or, in Ridley Scott's words, "retrofitted," which points both to the future and the past. It rains all the time. Ridley Scott explained in an interview: "The constant rain was 'dramatic glue,' if you like. It also amused me to think that it was taking place in Los Angeles, meaning the whole weather patterns would have changed by 2019. If L.A. gets all the rain, then maybe New York would get the sunshine" (51).

Blade Runner integrates futurist design associated with science fiction and film noir style and aesthetics. The interior design, for example in the opening, cites 1940s noir style, and one of the first two characters we encounter, Mr Hunter, is dressed according to 1940s noir, which leads some scholars and critics to call

continued

the film "future noir" (see for example Sammon). In addition, the machines with which the Voight–Kampff tests are conducted are simultaneously technologically advanced and look outdated. The futuristic city is shown primarily at night (similar to noir), emphasizing the neon light of the advertisements and the constant rain. The buildings themselves are kept in art deco design harking back to the early twentieth century. Interior spaces show a pastiche of design from the past and an imagined future. Noir is referenced not only in the interior design but also through cinematic lighting choices, for example, when Deckard visits his boss and walks in through a door with blinds that create the light-and-shadow effect typical of noir. Los Angeles also serves to connect noir to science fiction. The difference between inside and outside disappears when it rains inside the house. The consistent darkness, smokiness, and dreariness inside and outside create the general mood of the film.

Blade Runner represents an important juncture in the science-fiction genre. It takes place in the future but the cityscape shows decay. The film poses the problem of recognition for the audience: are we seeing and experiencing human subjectivity or not? One of the markers of humanness is affective memory. The use of photos to call up stored memories, self-referentially points to film as visual medium. Scott Bukatman claims that *Blade Runner* "is all about vision" (7). The problem of misrecognition of humans was introduced with the performance of the double Maria in *Metropolis*, played by the same actress, Brigitte Helm. The important turn in science fiction film connects the alienated cityscape, that no longer represents a futuristic, utopian, urban fantasy, to the question regarding the essence of human subjectivity.

The postmodern design concerns the pastiche not only of different periods but also of different cultures. The set design is reminiscent of ancient Egyptian architecture and design in the form of pyramids. The main characters are white, while Chinatown becomes the visual backdrop, and multiple languages, like German, make up the aural backdrop. Janet Staiger suggests that the postmodernism of the film finds its expression not only in the setting of the film but also in the multilingual advertisements: "*Blade Runner*, for instance, describes 'Cityspeak' as a combination of five languages, while advertising montages constructed from Japanese and English become ideograms spreading across Times Square-style billboards" (116). When Roy rescues Deckard he explains to him what it means to be a slave. In an American context, however, the discourse about slavery eerily points to the absence of African-Americans.

When Deckard visits the Tyrell Corporation, he meets Rachael and administers the Voight–Kampff test. Rachael embodies the look of the femme fatale of film noir

Figure 6.1 Ridley Scott. *Blade Runner* (1982): The postmodern, dark city

through her costume, her hair, and an acting style that does not express emotions. While Deckard administers the Voight–Kampff test, she is sitting in the dark, shot from one side, smoking. *Blade Runner* makes use of the conventions of film noir to mix genres and portray the future, but also to reinterpret the cycle of film noir by showing the cold exterior of female characters in film noir as robotic, since all female characters in *Blade Runner* are replicants.

The strong women, revealed to be replicants, die in very violent ways; in film noir female transgression was also punished. While Rachael, Deckard's love interest, is coded as from the 1940s, the other strong, female replicants are coded as punk through their costume, hair, and transgressive female behavior, which Sammon contextualizes with the interest of cyberpunk in the 1980s as a "subbranch of science fiction" (325). According to Sammon, it "embraced the diversity of the increasingly complex social, cultural, and scientific landscapes that decade produced, while simultaneously questioning the value system of the Reagan-era power structure" (325). Cyberpunk settings reflect the postmodern quality of the films discussed here: "set in a sprawling megalopolis of the near, dark, and decadent future, pitted hard-edged, street-level outlaws against omniscient

continued

(and corrupt) corporations, and viewed emerging hypertechnologies with equal portions of fascination and distrust" (325). *Blade Runner*'s political position, however, is ambivalent, since the characters that most clearly embody a sexualized version of punk are the two female replicants, Pris and Zhora, and both are killed easily, violently, and successfully by Deckard. The narrative punishes the women for their liberated sexual attitudes that are associated with punk subculture and reinterprets their free sexuality as man-made, quite similar to the construction of the sexualized Maria by a mad scientist.

Blade Runner had a consistent cult following, its appeal similar to that of *Metropolis*. Both films engage with the political questions of utopia and dystopia in their narrative but fetishize the cinematic potential of the spectacle of the city, enabled by technological sophistication. *Blade Runner*'s strategy of aesthetic pastiche, including the references to the modernist past, the retrofitting of the setting and design, the invocation of urban, cool film noir, and the integration of the urban subculture of punk, created an aesthetic shock that was embraced by an audience. That the film had such a power over its audiences speaks to the power of the image, particularly in the vision of a city.

Further reading

Scott Bukatman (1997) *Blade Runner*, London: British Film Institute. Good overview of approaches to *Blade Runner*, including discussions of visuality, cyber punk, and contrast to urban spaces in other science fictions, such as *Things to Come*.

Joshua Clover (2004) *The Matrix*, London: British Film Institute. Very convincing argument about the changes in labor in relationship to the "edge of construct film."

Annette Kuhn (ed.) (1999) *Alien Zone II: The Spaces of Science Fiction*, London: Verso. A collection of essays organized around different kinds of spatial formations in science fiction, including "city spaces."

Vivian Sobchack (2004) *Screening Space: The American Science Fiction Film*, New Brunswick, NJ: Rutgers University Press. The volume discusses spaces in science fiction, historically organized, with particular attention to science-fiction films as genre. It provides a context for the discussion advanced in this chapter, at least in clarifying that urban space is not a privileged space in science fiction.

Essential viewing

Jean-Luc Godard. *Alphaville* (1965)

Fritz Lang. *Metropolis* (1927)

William Cameron Menzies. *Things to Come* (1936)

Ridley Scott. *Blade Runner* (1982)

Andy and Larry Wachowski. *The Matrix* (1999)

SECTION III

While Section I brought together classic films in the canonical history of the city film from the 1920s to the 1970s and Section II employed a thematic approach in order to outline the development towards postmodernism, Section III addresses films organized around urban space and marginalized social groups. Urban life centrally shapes the experience of marginalized members of society, and in turn their existence defines contemporary cinematic portrayals of urban spaces. Chapter 7 discusses films about ghettos and barrios, Chapter 8 portrays sex and sexual minorities in the city, and Chapter 9 engages with the filmic representation of globalization and the city, often addressing issues of travel, migration, refugees, and illegal labor.

This section also offers a different methodological approach from the two preceding sections. Section I took its categories from film studies, using aesthetic and national categories to select and organize the films discussed, namely the Weimar street film, film noir, and the French New Wave. Section II employed a thematic approach to theoretical questions regarding history and politics in the shifts from national to transnational and modernist to postmodernist cinema. Section III relies on the social sciences to understand each chapter's topic and then moves into an exploration of its cinematic representation. All of the films discussed in this section emphasize urban topography and spatial division in relationship to social identity. The three final chapters emphasize contemporary films on the premise that foregrounding the urban experience of marginalized social groups provides important insights into the current understanding of urbanism and urban space.

The phenomenon of migration into the city also connects these final chapters. Ghettos and barrios are defined by spatial restrictions but often develop as a result of movement into ethnically or racially circumscribed areas of the city. Gays and lesbians have historically migrated to metropolitan areas because of the anonymity and comparative lack of social control that enables communities organized around non-normative sexual behavior. A chapter devoted to analyzing the filmic depiction of transnational migration into global cities across different national borders completes this section.

The films' production, narratives, and aesthetics mirror the survival strategies of the represented marginalized groups. Films on ghettos and barrios portray characters with limited means forced to recycle the discarded and pushed into illegality; they also reflect filming with limited resources and reliance on free or cheap labor. Films about gays and lesbians mimic the strategies developed to negotiate invisible and denounced sexual identities through the use of camp and cinematic codes that can be deciphered by members of the in-group. Films about globalization address different audiences around the world and circulate, like their characters, in the legal routes of global distribution on the one hand and illegal downloading and "bootleg" markets on the other.

Because these films negotiate marginalized identities in the urban context, they often dialogue with their respective communities. For example, in the past films about gays and lesbians circulated underground, but now gay and lesbian film festivals fulfill a crucial function for the reception and circulation of the films in the community. While all the films in question emerge from the experience of urban marginalization, they are also prone to the pressures of national and international markets. Therefore all three chapters address both independent cinema that captures and critically reflects on the material and psychological conditions of ghettoization in contrast to films made by the entertainment industry that fetishize, exploit, and circulate commodified images of the ghetto film, as well as gay and lesbian, and global cinema.

Finally, Section III concludes this book with a discussion of transnational cinema, demonstrating the shift from national cinema, as projected earlier. Chapter 9 includes only films that represent narratives of migration in relation to the global city, but even those films are sometimes based within national cinema. As Section I made clear with regard to national cinemas, there are neither purely national nor purely transnational cinemas, only degrees of differentiation on the continuum of film production and distribution. Transnational films comment on globalization, including the contemporary development of the global megalopolis.

Film as artistic expression not only represents but comments on and intervenes in the social reality of ghettos and barrios, the nexus between sexuality and the city, and the transformation of cities through globalization.

7 Ghettos and barrios

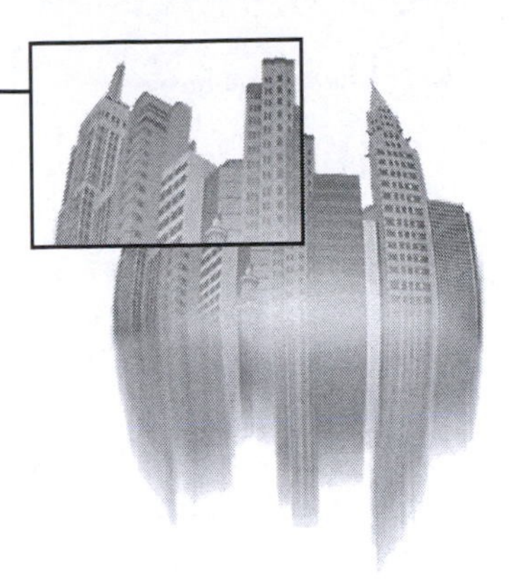

By the late 1960s, virtually all American cities with significant black populations had come to house large ghettos characterized by extreme segregation and spatial isolation.

Douglas S. Massey and Nancy A. Denton

People get tired because their daily life is so hopeless. You have to put some sort of entertainment out there. Let them live through your triumph vicariously. . .

Pam Grier

Learning objectives

- To gain insight into sociologically and historically based definitions of racial ghettos
- To comprehend the different ways in which studios and independent filmmakers capture the urban ghetto and enclave on film
- To outline the different kinds of spatial politics associated with ghettocentric films
- To capture the cinematic ghetto aesthetic in urban, national, and transnational contexts

Introduction

This chapter traces accounts of urban segregation and the constitution of ghettos in the industrial West as a framework for marginal film production. It illustrates how ghetto aesthetics in cinema share in larger cultural practices of recycling – the birth of hip hop, fashion, graffiti, urban murals – that have led to an infrastructure based on a continuum of semi-legal, illegal, and criminal activity of self-fashioned recirculating of goods and refashioning of images. These kinds of representations can be traced from mid-twentieth-century Italian neorealism, via the African-American Blaxploitation boom of the late 1960s and the politically inflected, Afrocentric, independent, art-house cinema, to Spike Lee's independent cinema and the mainstreaming of "the ghetto film," the birth of which the US saw with John Singleton's *Boyz N the Hood* (1991). The mainstream success of this genre, which paralleled the transnational crossover success of black hip hop music, led to an increased production within the US and a transnationalization of a formulaic genre expressed in such films as Brazilian Fernando Meirelles's *City of God* (2002) and South African Gavin Hood's *Tsotsi* (2005). These mainstreamed, high production-value, ghetto films appropriate and fashion a ghetto style and generic narrative formula that are detached from any political analysis of urban reality offered by urban studies and the lived experience of ghettos, barrios, and ethnic neighborhoods. Beyond this contemporary transnational and national appropriation that exploits fantasies of poverty for action-filled stories, however, ghettocentric representations by and about marginalized immigrants and other minority communities – be it French-Africans, Afro-British, or Turkish-Germans – have emerged from around the world.

The timeline in this chapter is different from earlier chapters in that it relies on the more famous films to trace hidden influences. I begin with an early paradigmatic account of the ghetto by W.E.B. Du Bois in order to arrive at an important moment in the US, the early 1990s, when it saw the birth of the contemporary, urban, ghetto film which laid the foundation for the transnational genre as we know it now. I trace the influences of the contemporary genre back to distinct representational phenomena in the 1970s: Blaxploitation and independent Black Cinema in the US. The latter then leads us back in history to the postwar period in Italy and the development of Italian neorealism, with two examples from the early 1960s. The chapter ends with a summary of contemporary variations of films about the ghetto and a case study of *The Harder They Come* (Perry Henzell, 1972), an important Jamaican film which recycles American and national myths and genre conventions in a setting that takes its political impetus from its authentic portrayal of Jamaica (see Warner). To avoid confusion in tracing influences forwards and backwards in history, years have been added in front of the section headings.

1890: the spatialization of the "negro problem"

Even though the focus here is on cinematic representations of urban ghettoization in the late twentieth century, neither the social discussion of living conditions – the spatial limitations, and the class, racial, and ethnic dimensions of ghettoization – nor its cinematic representation belongs only to the twentieth century. W.E.B. Du Bois's essay "The Negro Problem of Philadelphia," published in this book *The Philadelphia Negro*, written in the 1890s while he was living in the African-

American neighborhood of Philadelphia, is a paradigmatic account of the racial ghetto in the late nineteenth century, which emphasizes diversity from within. The quest to understand the "Negro problem" through "a study of Negroes in the Seventh Ward, the city's Black ghetto," illustrates the conflation of American minorities with the perceived problems of urban blight (119).

Du Bois saw the ghetto as "a city within a city" whose members "do not form an integral part of the larger social group," and proposed that this is not an unusual phenomenon, since there are other unassimilated ethnic groups in the city, such as Jews and Italians (120). But he modified his position that degrees of segregation are normal by emphasizing the "conspicuous" segregation of African-Americans (120). Du Bois's account gives much description of particular streets and their surroundings, the conditions and the material of the houses, and the histories of specific riots. He aligns individuals with various streets, alleys, houses, and apartments, according to profession and income. He also differentiates the level of noise and criminal and violent activity, having in mind the stereotypes of ghetto inhabitants. However, in contrast to other studies, these stereotypes were not his primary concern, but rather the effect of discrimination on the opportunities and choices in terms of labor, and the money that can flow into the neighborhoods in question. Du Bois's study continues to be important today because it includes a subtle account of the diversity within the ghetto.

1991: "the new ghetto aesthetic"

A century later, in the early 1990s, the ghetto suddenly had its moment in the limelight in a wave of highly successful crossover films. Jacquie Jones's "The New Ghetto Aesthetic" observes a new popular phenomenon at the time, the ghetto film, and makes a critical intervention in its representational politics. At the end of 1991, Jones takes stock of a slew of films by African-American directors that take place in the "contemporary urban ghetto" and concludes: "Only one . . . contains a valuable leading role for a Black female actor. And none are directed by Black women" (32). Her essay provides a critical stocktaking of the explosion in new Black film, which according to her follows Hollywood conventions and therefore neither undermines existing racial or gender politics. Instead, Jones interprets this new output as a result of "its marketability," because "Hollywood hopes to capitalize on the success of recent low-budget Black-made films" in creating "a battery of films which 'illuminate' the life of the young Black male, the nation's most recent sociological curiosity" (33).

Jones implies here that these films cater to a voyeuristic audience simultaneously attracted to and repulsed by a fantasy about the ghetto as a taboo zone in need of explication and translation. She contrasts the economic successes of the time, such as Mario Van Peebles's *New Jack City* (1991), which shows Harlem's "crime-infested Black ghetto," to the engagement of his father, Melvin Van Peebles, in "guerilla" filmmaking: "Yet while the elder Van Peebles envisioned the law as necessarily an enemy of Black people, the younger positions it as a savior to Black people in communities fraught with self-imposed lawlessness" (34). And she contrasts the audience and financial success of *New Jack City* and *Boyz N the Hood* with low-budget film indebted to independent Black cinema that could not find distributors. Jones describes the American urban ghetto films that show the ghetto as masculinist, undifferentiated, and utterly violent, and turn the ghetto into a commodity that feeds liberal racist and sexist anxieties and desires of mainstream America.

1991 became *the* year of the new American urban ghetto film when John Singleton's *Boyz N the Hood* (1991) and Mario Van Peebles's *New Jack City* (1991) hit the big screen. Both tell stories of violence among urban Blacks associated with drugs and gang warfare and are set in decaying urban locales made to represent such problems as policing, drugs, gentrification; lack of jobs, resources, and education; incarceration and gang warfare. *Boyz N the Hood* takes place in south-central Los Angeles, and *New Jack City* in New York City. Their similarities emerge out of the discourse on urbanity – geopolicital, urban, and industrial developments – in the late 1980s. Their differences concern the genres they reference, the predecessors they evoke and disavow, and the paths that they prepare for the development of the genre.

New Jack City, which came out first, presents the story of violent conflict between a gang of urban drug-dealers and the police and is told in conventional action-film style and narrative. High-profile actors and actresses, including beauty queens and rap artists such as Vanessa Williams, Wesley Snipes, Ice T, and Chris Rock, add recognizable Hollywood glamor to the film's claim to realist urban grittiness. *New Jack City* portrays the rise of gangster Nino Brown who takes over a house in New York City for crack production and distribution. He is brought down by a multicultural group within the police force, including Scotty Appleton (Ice T), Stone (Mario Van Peebles), Nick Peretti (Judd Jelson), and their infiltrator from the street, former drug addict Pookie (Chris Rock). This film shares with *Boyz N the Hood* the claim to a realist depiction of the urban space defined by poverty, violence, and drugs; both films claim realism by using documentary audio footage in their openings. In *New Jack City* we hear fragments of voices from the radio over a shot of New York City accompanied by hip hop music: "income for the rich/ homelessness is at an all-time high/ the three were victims of a drug deal gone bad." *Boyz N the Hood* opens with a similar audio collage: fragments of a conversation that begins with a shooting are heard while the screen is still black.

Then a statement appears on the screen – "One out of every twenty-one Black American males will be murdered in their lifetime" – followed by a police radio fragment – "officer needs assistance . . . at the corner of Crenshaw and . . ." – and a small child's voice: "They shot my brother." This prelude ends with the sociological and factual indictment written on the screen: "Most will die at the hands of another Black male."

Each film ends with a claim to documentary realism with a text that scrolls across the screen. *New Jack City*'s text reads: "Although this is a fictional story, there are Nino Browns in every major city in America. If we don't confront the problem realistically – without empty slogans and promises – then drugs will continue to destroy our country." *Boyz N the Hood* also associates its claim to truth with a written text at the film's end to provide seemingly factual information about fictional characters: "The next day Doughboy saw his brother buried./ Two weeks later he was murdered./ In the fall Tre went to Morehouse College, in Atlanta, Georgia./ With Brandi across the way at Spelman College." These paradigmatic claims to authenticity framing the films picture the ghetto as a space from which real stories emerge, stories which correspond to sociological and ethnographic accounts that objectify its inhabitants. As a consequence, however, such films also purport to show the truth about the ghetto and leave viewers with the illusion of a privileged knowledge of the parts of America that are simultaneously invisible and hypervisible.

The American ghetto films from the early 1990s reflect urban changes in the late 1980s. William Julius Wilson describes the intensification of poverty during the early 1990s in American inner cities in contrast to the post-Second World War period:

> In 1959, less than one-third of the poverty population in the United States lived in metropolitan central cities. By 1991, the central cities included close to half of the nation's poor. Many of the most rapid increases in concentrated poverty have occurred in African-American neighborhoods. (129)

According to Wilson the exodus of non-poor from Black neighborhoods that had been mixed-income areas (as described by W.E.B. Du Bois) created the extreme poverty in post-segregation American inner city ghettos (131). Wilson faults "the rapid growth of joblessness" for the "new urban poverty" (132), which created the new urban ghetto as "the jobless ghetto, which features a severe lack of basic opportunities and resources, and inadequate social controls" (134). The emergence of the new ghetto film in the early 1990s reflected those changes by showing urban blight and the experience of discrimination. But the genre also inscribes a limited perspective as truth and thus creates a commodified product for the film industry.

Representations of violence, misogyny, and racist stereotypes came to dominate the genre and contributed to its economic success. Films that offered differentiated accounts of the economic and psychological causes and effects of poverty and ghettoization were marginalized. In order to map out the commodification and resistance associated with the ghetto film, I compare and contrast *New Jack City* and *Boyz N the Hood* because they show characteristics of what became the "ghettocentric hood movie," though they are also distinct from each other in important ways. Both take place in a clearly circumscribed urban locale.

New Jack City opens with a thug holding a white man off a bridge and then dropping him in the water, a moment that is not integrated into the narrative. A text on the screen announces that this is "The City 1986." The ghetto in *New Jack City* is less a circumscribed location than a style with potential for commodification. The very first shot of the main group of characters shows them with Kangol hats, a branding that reappears in Quentin Tarantino's *Jackie Brown* (1997). Kangol hats were first brought out in 1938 by French maker of berets Jakob Henryk Spriergeren. Throughout the twentieth century it was the company's advertisement method to be associated with stars, from those of the British Olympics in 1948 to Greta Garbo and Marlene Dietrich. In 1984 rapper LL Cool J wore a Kangol hat on his album *Radio* to create street credibility as image for the brand. In addition to the Kangol hat, the main characters representing "the ghetto" wear black leather and big gold chains. Their bond is an affair among men based on brotherhood. The three main characters drive down the street and discuss the new drug "freebase." When Nino says about The Carter, the building the gang is going to take over, "It is like Beirut, they become hostages, we will own this fucking city," the film pays lip-service to a political position that it does not live up to and claims a political context, even though it is determined only by its action narrative. The supposed political context is the continuum, suggested by Graham (see Chapter 5) of the city at war, in civil war, or divided by racist and ethnic conflicts.

New Jack City situates itself in the tradition of the ethnic gangster film: repeatedly the characters watch Brian de Palma's *Scarface* (1983), the ur-text of the ethnic gangster film. *New Jack City* ties violence to sex, as when Gee Money's girlfriend betrays him for his boss, the gangster Nino (parallel to the narrative of *Scarface*) and, in order to seduce Nino, performs a striptease in front of an overblown *Scarface* image. Nino fails because of his hyperbolic desire for power and the betrayal of the male brotherhood. In the end, the gang members are all killed and Scotty, the African-American cop, realizes that his mother was killed by Nino and has the moral imperative to connect retributary violence with masculinist sexuality when he threatens Nino: "This ain't business, bitch, this is personal. I want to kill you so bad, my dick's hard." The film ends with the invocation of politics, when Nino explains in court that white people are bringing the drugs into the ghetto: "Do you see a poppyfield in Harlem?" Even though politics is invoked at the film's opening and ending, the film offers no analysis, cinematic innovation, redress, or solution beyond brute violence.

Boyz N the Hood is more subtle and critical and characterizes the ghetto more carefully than the exaggerated, action-film stunts of *New Jack City*. When the film's main character Tre is in his teens, his mother takes him to his dad, Furious, who lives in the ghetto, to raise him. Furious raises Tre with strict morals, including a clear notion of what it means to be a Black man. Furious is contrasted with Tre's mother, who is getting an education and does not want to take care of her son. Her education, however, is undefined and marks her as a selfish mother who rather does something for herself than take care of her son. In contrast to her, both Tre and his love interest, Brandi, leave for historically Black colleges in the South at the end of the film. The group of main characters is rounded out by Tre's neighborhood friends, particularly Doughboy and Chris who live across the street with their mother. The boys have different fathers; the mother, however, favours Chris, who is supposed to go to college on a football scholarship. The story of the group of friends begins to unfold after a scene in which they meet at a backyard barbecue celebrating Doughboy's release from jail. Doughboy has run-ins with members of local gangs who end up killing his brother Chris. Doughboy avenges Chris's death by killing those responsible. The configuration of the single but moral father Furious and the single but immoral mother of Doughboy and Chris predetermines the development of the characters: Doughboy and Chris will die violently in the ghetto, while Tre and Brandi will be able to leave.

The film opens with a scene from the perspective of the main characters as children, articulating a different point of view from that of the exoticizing, scandalous perspective on the ghetto in *New Jack City*. The *mise-en-scène* marks the space as ghetto with details of an iron fence in front of a house, a shopping cart left in the street, loose dogs running around, and a child waiting alone on a street corner. Symbolically the film begins with a stop sign and a plane flying over it, marking both the immediate local and the possible global connections. The stop sign visually signifies that movement is limited, while the plane signals the impossible possibility of escape. These are the two, individual, possible responses to ghettoization announced by the opening and they reflect the two narrative trajectories of the ghetto film in general.

Whereas *New Jack City* invokes the voyeurism of the spectator with acts of senseless violence at the outset, the opening of *Boyz N the Hood* shows us four children witnessing a dead body for the first time, a murder victim left behind in the ghetto by an uncaring bureaucracy. Similar to *New Jack City*, *Boyz N the Hood* is framed by political commentary on the racism of policing and of education. A white teacher lectures the class about the Puritans and Thanksgiving, after which little Tre gives an Afrocentric lecture.

Boyz N the Hood takes place in Crenshaw, Los Angeles. The only character who leaves this circumscribed environment is the Afrocentric father Furious, who we see at the ocean early on. While the opening with children suggests a tale of socialization in the ghetto, the film creates a prescriptive narrative about those who do well and those who end badly, for example when Doughboy and Chris turn into juvenile delinquents. The film criticizes the violence but at the same time violent shootings motivate and punctuate the narrative. Cinematically the film fetishizes the repeated violent outbursts by showing them in slow motion. Despite its subtlety in comparison to *New Jack City*, *Boyz N the Hood* created a formula for the ghetto film and paved the way for a genre that mobilizes stereotypical portrayals of violence, ghetto style, hypermasculinity, and pathologically excessive female sexuality.

Jones analyzes the female figures in the new ghetto-centered films as "bitches" and "hos," downplaying Furious as a continuation of the politically enlightened figures of the 1970s films by Haile Gerima and Charles Burnett that I will discuss shortly. The sexualized female figure central to the commodification of the ghetto film finds its ultimate recent expression in the "pimp and ho" film by Craig Brewer titled *Hustle & Flow* (2005), in which all the female characters are prostitutes surrounding the main character who is a pimp and hip hop artist.

Todd Williams's *Friendly Fire: Making an Urban Legend* (2003) documents the making of *Boyz N the Hood* and shores up the authenticity of the film's ghetto representation. The documentary takes as its point of departure the moment in *Boyz N the Hood* when Furious instructs the teenagers Ricky and Tre about the politics of gentrification, which serves as the political centerpiece of the film. Because this is the only excerpt from the film included in the documentary and because it opens the documentary, it also serves as its motto: "Why is it that there is a gunshop on every corner?" The documentary thus offers both itself and the feature film as a political answer to the question. As in hip hop and in cinematic strategies, ghetto-centered discourse established post-1970s claims to authenticity with one eye on the market. After the shot of Furious's speech about gentrification in the ghetto, we are informed by John Singleton, "You know, that was our life," a sentiment later echoed by Ice Cube (Doughboy): "This is my hood; this is our life." We then learn about the 1980s and early 1990s, when "LA was like a police state," the context of the gang wars of the Cribs and the Bloods, and the "enormous tension between the police and the neighborhood." Singleton returned to the street in the south-central neighborhood of Los Angeles where he grew up – Lawrence Street – and this is the first street sign we see in the film.

The documentary emphasizes the success of *Boyz N the Hood*, beginning with the fact that at 22 John Singleton was the youngest person to be nominated for an Oscar and the first African-American nominated for an Oscar for directing. The actors in the documentary acknowledge that the film launched their careers and that it was the most profitable movie of its year. In order for Todd Williams, the director of the documentary, and John Singleton and the cast to be able to create what they call "an urban legend" about the making of *Boyz N the Hood*, they have to disavow their cinematic roots. They mention only Spike Lee as a predecessor in Black filmmaking. Cast members repeatedly claim, as does Cuba Gooding Junior: "This was the first time it was a predominantly Black crew."

Gooding's statement can serve as a springboard to investigate the unacknowledged forerunners of *Boyz In the Hood* because in fact there is a history of films produced and directed by African-Americans with all-African-American crews and casts situated "in the ghetto." During the segregated film industries of the 1920s, Oscar Micheaux headed an all-Black film company and wrote, directed, produced, and distributed films representing the Black community to a segregated Black audience. Then the Black independent cinema movement of the 1970s included Haile Gerima and Charles Burnett who directed films about "the ghetto" with all-Black casts and production teams. The Blaxploitaton boom of the same period created action-filled ghetto narratives that were directed primarily by Black directors, located in the ghetto, shown primarily in urban B-movie houses. After their initial success they were also produced by Hollywood studios; hence Blaxploitation has a curious position between independence and the studio system. The contemporary ghetto film integrates aspects of both film movements: the documentary realism of a politically independent cinema movement, and the fantastic, hyperbolic, masculinist violence and sexuality of Blaxploitation.

1971–73: ghetto fabulousness

Blaxploitation denotes low-budget action films located in urban environments centered on hyperbolic Black heroes. The low-budget films emphasized action and exaggerated narratives because directors did not have access to the studio system's high-quality *mise-en-scène*, costume, and editing. Melvin Van Peebles independently produced and directed *Sweet Sweetback's Baadasssss Song* (1971) with an all-Black crew and cast. The visible low budget and the circulation in inner-city B-movie theaters result from a lack of resources, which in turn mirrors the conditions of the ghetto in the film's cinematic practices. Once the studios realized the potential for the success of Blaxploitation, however, they mimicked the low-budget quality and exploited the conventions of sexualized and violent representation of African-Americans.

Blaxploitation film *Black Caesar* (1973), by Larry Cohen, follows the *Scarface* gangster narrative formula within a ghetto narrative. Tommy, the fatherless son of a Black maid, had been abused by a white policeman as a child and rises as a gangster until he is betrayed by his woman and his best friend from childhood. The story is set in Manhattan, New York City, and much of the narrative is determined by the characters' movement through the city. Tommy's original home is a dilapidated building, to which he returns twice, once when his absent father reappears and again at the very end of the film, after he has been shot repeatedly. Tommy's hope to rise in the mob is challenged when one of the mobsters says to him, "They'll never accept a nigger in the syndicate," and he answers, "Everybody is a liberal today, accept it." His meteoric rise as a gangster is due to his "smarts," with which he is able to buy his white lawyer's apartment, including the maid who is revealed to be his mother. The film offers a fantasy of hyperbolic success through illegal activity, but portrays the escape from the ghetto as impossible, since it reproduces itself through senseless violence. In contrast to the new ghetto film, however, *Black Caesar* is motivated by a primal scene of humiliating and violent racism that motivates the narrative and invites viewers' identification.

Gordon Parks's well-known *Shaft* (1971) has given us one of the most famous icons of Black masculinity in the figure of Shaft, who embodies fetishized Black tough masculinity and claims an iconic status without past, childhood, roots, or origins. His figure marks the transition from a gangster narrative that relies on a biography tied to the ghetto as an explanatory model to the articulation of ghetto style, which enables him to negotiate the spaces of the city. Thus Shaft, famous for his self-confident walk through the city, his stylish 1970s leather coats and tight pants, his constant readiness for and satisfaction with sex, his smarts and his foresight, seamlessly moves through the different sections and social groups of the city: uptown and downtown Manhattan, the police, the black mob, and the Italian mob, as well as different places in the city – coffee-houses, restaurants, police stations, bars, stores, various apartments. In this studio production, the ghetto becomes one urban space that can be negotiated by those black men who incorporate the knowledge, language, and skills required to successfully cross the invisible class, ethnic, gender, and racial boundaries that structure the topography of a city. The film employs the main character to redefine Blackness as urban and urban Blackness as cool by invoking the style and skill that emerges from the fantasy of the ghetto that ignores its economic and geographical limitation and relies on older stereotypes of black hypermasculinity. Shaft's walk through the city is accompanied by the Isaac Hayes song *A Sex Machine: Shaft*, and the fantasy of being the walking Shaft offers a cross-racial, crossover male fantasy of confident ownership of the urban terrain that is not rooted in a specific individual or collective history or geography. Shaft's sexual prowess extends to women and the city, conflating the two. Racial discrimination becomes the content of witty exchanges, as when his girlfriend calls him on the phone and asks him whether he has a problem. Shaft says: "I have a couple of them. I was born black. I was born poor," and his girlfriend responds with a complete non-*sequitur*: "I love you." Racial discrimination and class stratification are turned into pretexts for erotic encounters.

Blaxploitation did not advance merely a masculinist ideal: one of the genre's most famous stars is Pam Grier, who integrated sex appeal with a physical ability to fight. Jack Hill's *Coffy* (1973) offers an entirely different urban narrative from those discussed so far. Grier plays Coffy, a nurse whose sister is a drug addict, which motivates Coffy to take on and kill an entire narcotics organization using her sexuality and skills to escape seemingly impossible situations. In one scene Coffy hides razor blades within her big afro before a catfight, so that the other women cut their hands. The scene incorporates an acknowledgment of female ingenuity with the exploitation aspect of the genre, found in the staging of female catfights. The exaggerated action and narrative create the pleasure that comes from the film's hyperbolic representation. The emblematic quality of this scene for the hyperbolic politics of Blaxploitation can be seen in the citations of it in later films, including the Blaxploitation spoof *Undercover Brother* (Malcolm D. Lee, 2002) and the art-house meditation on identification and desire in relationship to Pam Grier as an icon of a strong female character in Etang Inyang's *Badass Supermama* (1996). At the end of the film, Coffy kills her lover, a congressman who articulates an empty political position, offering a fantasy of militant response to political corruption. Pam Grier embodies the Blaxploitation fantasy of an individual's revenge on forces of corruption and systematic discrimination.

While Blaxploitation offers a fantasy of urban sexuality and violence in which the underdog will always win hyperbolically, the parallel movement of the independent art-house cinema developed a juxtaposed artistic portrayal of the ghetto based on realism. By organizing its production as a communal experience of working together and sharing resources, the production mirrored the more productive aspects within communities with limited resources in its own collective organization of production. The fantastic urban heroes and heroines who were the stars of Blaxploitation and embodied its sexualized version of urban Blackness offered these fantasies primarily to an urban underclass. The pleasure of those films emerged from their hyperbolic action and their revenge narratives, not from a claim to realism.

1976–1977: beauty and struggle

Charles Burnett's *Killer of Sheep* (1977) takes place in south-central Los Angeles and tells the story of an all-Black environment determined by poverty, but its look is entirely and strikingly different from that of either Blaxploitation or the contemporary ghetto film. The extreme close-ups and the absence of establishing shots create a space that is defined by relationships and labor, and poverty is invoked but also permanently redeemed through the cinematic beauty of the stark and gritty but beautifully composed black-and-white shots. The urban spaces that Burnett creates cinematically eschew violence, destruction, limitation, and the breaking down of social structures. Instead, *Killer of Sheep* follows the emotional toil and the pleasure of everyday life, including showing children playing, throwing stones at each other, running, singing in the playground, dressing up, riding bikes, and counting while making handstands against house walls. The many scenes of children playing outside and inside the house are interwoven with snippets of the everyday life of the main nuclear family, the husband Sam, his wife, and their son and daughter. Their daily activities include house repairs, trying to buy a used engine to replace the one in the car, sitting at the kitchen table drinking coffee, playing dominoes, relaxing in the living room, and working.

With only sparse dialogue, we are brought to realize the wife's strong desire for her husband Sam, a desire that he does not return, and her loneliness within the family. There is little focus on character motivation. We see Sam work to make ends meet, sometimes with and sometimes without success. Towards the end, the group of friends takes a trip to Los Alameda, and on the way they have a flat tire. The banality of it all encapsulates the reality of living with limited means and the everyday strategies of survival more effectively than the extreme binaries of the ghetto film, such as life–death, escape–entrapment, success–defeat, and legal–illegal. When the characters return from their shortened day-trip, we witness a rare scene of affection between husband and wife, and thus, even though the group did not make it to the intended destination of the journey, the outing had a subtle affect on their relationship.

This film relies on extensive footage from Stan's workplace, a meat factory, which gives the film its title, *Killer of Sheep*. Like the images of home and street life, the scenes at work are strikingly beautiful despite their reliance on documentary conventions. Children's play and domestic scenes are repeatedly intercut with the men at work or just the sheep. While explicit politics are absent, the politics of representation communicate an understanding of lives that are characterized by limited choices but not determined by those limitations. Instead, the characters create their everyday lives out of those limitations day after day. Burnett shares this politics of representation with his fellow filmmaker Haile Gerima, his predecessors

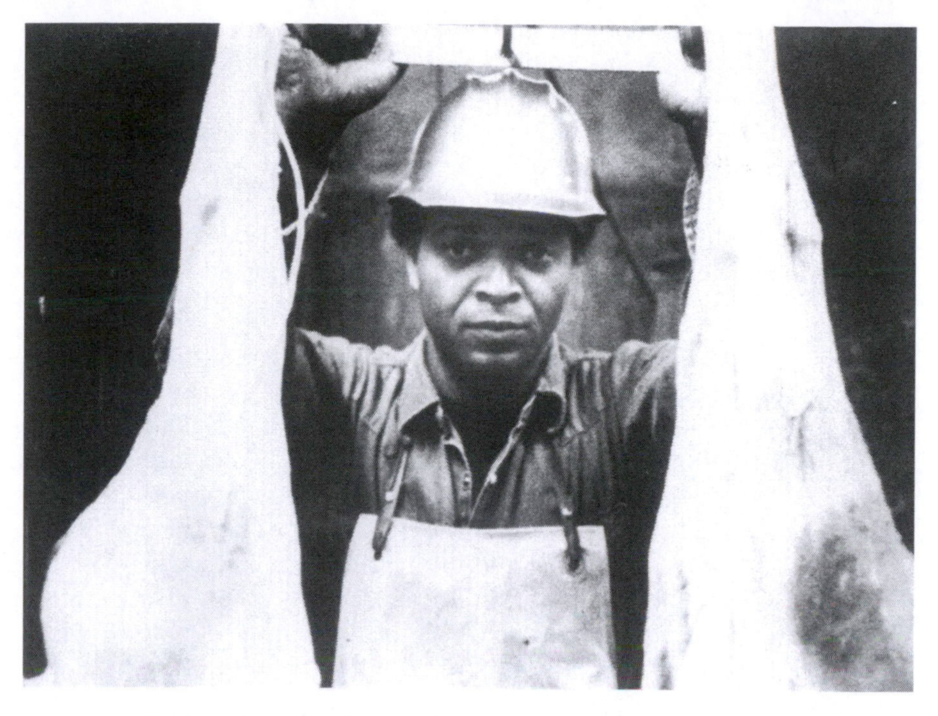

Figure 7.1 Charles Burnett. *Killer of Sheep* (1977): The aesthetics of the everyday

of Italian neorealism (1943–1962), and contemporary minority filmmakers in Europe and elsewhere who engage in similar strategies to portray socio-economic and psychological marginalization in urban environments.

In contrast to the contemporary ghetto film, 1970s independent cinema that addressed ghetto life often focused on female characters and portrayed violence in an specifically political context, witness Gerima's *Bush Mama* (1976). Like *Cléo from 5 to 7* it offers an unusual representation of a woman, the main character Dorothy, walking through the city, as the refrain of the narrative. Dorothy attempts to survive under conditions that are significantly harsher than Burnett's portrayal in *Killer of Sheep* on the one hand and simultaneously more benign, institutional, symptomatic, and systematic than the new ghetto aesthetics portrays them to be on the other. Her male lover, TC, is incarcerated and she raises her teenage daughter, Louann, alone. Dorothy is pregnant, and her social worker pressures her to have an abortion. She drinks in bars and discusses her addiction in self-help groups. Towards the end, she is arrested, her fetus is beaten out of her, her daughter is picked up by the police and raped, and Dorothy then kills the police officer. Dorothy and TC write each other letters that are read in voice-over, and Dorothy learns about radical politics from Louann's friend.

Instead of a linear narrative, the film captures fragments that create the space of the ghetto filled with memories, desires, and dialogues. We encounter characters inhabiting street corners, bars, bus stations, and unnamed meeting-places who watch and comment on the conditions of life. For example, at a bus-stop an unnamed character soliquizes on "the restaurant in nigger town, well-done piece of meat," and calls himself "the Prince of Homey." Not all characters have a function for the narrative; instead they negotiate the conditions within the frame of possibilities of a space that is institutionally circumscribed.

In *Bush Mama* the violence is systemically created by the law and its institutions. Like *New Jack City* and *Boyz N the Hood*, the film begins with the audio of police interviews over a slow-motion police arrest. *Bush Mama* continues the strategy of separating the audio from the image throughout the film and thus uncouples the institutional discourse from the dreams, fantasies, and unspoken aspirations of the characters. When Dorothy is lying on the sofa daydreaming, we hear voices asking questions characteristic of officialdom: Do you have a phone? Do you have a number? You have one daughter? Another child? The institutional discourse that pathologizes poor, Black women haunts Dorothy, even in her private space; obsessed with her body, these institutional practices are finally carried out in direct violence on her body.

Dorothy wanders the urban streets. But she is not a female version of Benjamin's leisurely wandering male *flâneur* who is beckoned by the phantasmagoria of the city. Instead she inhabits the public space because she is haunted by institutional pressures and lack of resources. We see Dorothy walking but never arriving anywhere; we are witnesses to the confined urbanity defined by racism. Whereas Cléo in *Cléo from 5 to 7* becomes increasingly free in the city, Dorothy becomes increasingly hurt and entrapped. At the end, beaten down physically, she writes a letter to TC which shows that she has nevertheless acquired a subjective voice: "Laws that protect people who have money . . . to read and to study . . . we all have to so that we can change it . . . talking to each other is not easy . . . most of the time I don't understand your letters . . . it's not easy . . . the idea is to win over more people like me . . . all day and all night when I eat and sleep . . . TC . . . the weight was off my head . . . TC . . . I love you . . . Dorothy." Both *Killer of Sheep* and *Bush Mama* offer sensitive and subtle accounts of the psychological and emotional damage done to individuals through the lack of resources. Gerima shows the dialectical relationship between political and institutional structures and private desires and affections. The love between TC and Dorothy is shaped by their experience of societal limitations, while their desire for each other also leads them to investigate and critique the politics around them. The spatial and economic limitation of the ghetto is visually reproduced through the repetition of Dorothy's walking and waiting in the streets. Narratively those spatial restrictions are also recreated through the motif of incarceration that leads to the split of the visual from the audible fabric of the film.

1961–1962: Italian neorealism – walking through Rome's ruins

The aesthetic realism in the portrayal of urban poverty in *Killer of Sheep* and *Bush Mama* invokes Italian neorealism, a movement from the early 1940s that focused on films set among the poor and working class in Italy, shot in black-and-white on location, using lay actors. Pier Paolo Pasolini's *Accattone* (1961) and *Mamma Roma* (1962) portray urban poverty in on-location shooting that captures the gritty reality of marginal lives on the outskirts of Rome. *Accattone* shows the transition from poor rural housing to the modern housing projects on the outskirts of Rome through the story of a pimp. *Mamma Roma* is similarly set in the housing projects around Rome where Mamma Roma, an aging prostitute, is raising her son and desperately trying to prevent him from becoming a criminal.

Figure 7.2 Pier Paolo Pasolini. *Accattone* (1961): Walking through the construction sites in Rome's outskirts

The neo-realist films *Accattone* and *Mamma Roma* reflect class organization in the city of Rome, which differs from the dynamics of cities in the United States. The urban poor are not contained in the inner city, but in new buildings built on the periphery of the metropolis during the early 1960s. The majority of shots in *Accattone* and *Mamma Roma* therefore show the no-man's-land between the new and impersonal housing construction and the empty spaces surrounding the buildings, which include some ruins. In *Mamma Roma* the modern architecture and the ruins create an in-between space that is reminiscent of the "desert of the real" that is shown to Neo in *The Matrix* and the post-apocalyptic landscape in *Things to Come*, but results from urban planning and housing construction that do not integrate the past with the present. These voids create the spaces where the youth of the area come together, and where Mamma Roma's son develops into a thief who by the end of the film is under arrest. In contrast to him, Mamma Roma belongs to the past of Rome, walking through the city, integrating what feminists have analyzed as the whore–virgin mother dichotomy. She sacrifices herself for her son and does not want him to find out that she is a prostitute, but makes herself vulnerable to blackmail. One important scene shows her walking the streets of Rome at night, telling a story, while different men drift in and out of her company and the cinematic frame. In the old part of the city, Mamma Roma is a storyteller, in touch with her own and Rome's past. But in the new construction site, in a land for migrants and nomads where organic ties are destroyed and ruins have lost their meaning, she cannot protect her son.

In the end, her son is arrested by the police and tied to a table. Haunted by nightmares, he wants to return to the country, cries for his mother, and feels sick. The shots of him in prison are intercut with an establishing shot of the housing complex, tying his personal development to the urban development of Rome. While he is suffering, Mamma Roma is looking out of the window, similar to Dorothy in *Bush Mama*. She sees the landscape of ruins that mirrors the interior brokenness and displacement of the characters. When they tell her at the market that her son is dead, she runs back to her apartment and tries to kill herself. The community of people holds her back but cannot offer her any hope. The film ends with a shot of the buildings, situating her individual fate as a parable for a community lost and alienated in the postwar shift from rural to urban migration.

1968–2005: from then to now – the ghetto film goes global

The discourse on ghettoization is entrenched with class and racial stratification but it is also linked with migration in its two common forms of rural to urban and from one nation to another. The migration from the countryside to the city and the settlement in limited and poor areas there is also portrayed in other national cinemas, for example in the Turkish film *The Horse* (Ali Özgentürk, 1982). A father and son are moving to the city, and the father acquires a cart, which he pushes through the city of Istanbul. Attractive images of the city, with its famous mosques, contrast with the limited space the father and son can experience, until finally the father dies in an altercation and the son has to return home alone. The use of lay actors and an emphasis on young men in the public sphere also characterize early American films about ethnic enclaves in such films as Martin Scorsese's *Who's That Knocking on my Door?* (1968). These have in turn been recycled in three contemporary transnational versions depicting the urban minority ghetto: first, the immigration comedy, such as Damien O'Donnell's *East Is East* (1999); second, the films in the tradition of Italian neorealism, such as Thomas Arslan's *Brothers and Sisters* (1997); and, third, the transnational ghetto blockbuster, such as Fernando Meirelles's *City of God* (2003) and Gavin Hood's *Tsotsi* (2005).

In the US, the ghetto-centered action film has become a formula used as a vehicle primarily for rap stars, witness Jim Sheridan's *Get Rich or Die Tryin'* (2005), in which the rap artist 50 Cent claims to tell his rags-to-riches story. The hyperbolic ghetto fabulousness which exaggerates the *accoutrements* associated with ghetto style figures, such as "the pimp" or "the ho," can be traced back to Blaxploitation, while the socially critical realism can in turn be traced back to the independent Black cinema of Gerima and Burnett, which in turn echoes Italian neorealism. Because Hollywood dominates the international market, films like *City of God* and *Tsotsi* from the national cinemas of Brazil and South Africa can be economically successful internationally by mimicking the American urban ghetto film that exploits the representation of poverty. The mass distribution of Third World films fetishizing the violence of the ghetto at the cost of low-budget films that more subtly address issues of urban poverty not only creates a misconception about the urban poor in specific Third World countries, but also lends itself to seeing entire Third World countries as ghettos. Several classic Third World films have focused on an inside look at slums and ghettos, including Mira Nair's *Salaam Bombay!* (1988), Euzhan Palcy's *Sugarcane Alley* (Martinique, 1983), Marcel Camus's *Black Orpheus* (Brazil, 1959), Luis Buñuel's *Los Olvidados* (Mexico, 1950), and *The Horse* (Turkey). The case study which follows shows how one film from the small Jamaican national film industry can, however, successfully negotiate this complex national and international terrain.

Case Study 7 **Perry Henzell's *The Harder They Come* (1973)**

Most of the films discussed in this chapter were made in the US or Europe, where the ghetto, the barrio, and the ethnic neighborhood constitute anomalies within industrialized cities. The chapter has focused on films and film movements from the 1960s to the present, films that negotiate the reality of the ghetto in the context of First World capitalism. This case study focuses on possibly the most influential of Jamaican films and one of the most important films from the Caribbean, Perry Henzell's *The Harder They Come* (1973), which shares characteristics with some of the films discussed in this chapter. *The Harder They Come*, however, also advances and confronts politics from the perspective of a Third World country in its negotiation of urban and class politics, the effects of colonial history, and global economic exploitation. In addition, the conditions for filmmaking in the Third World in general, but in the Caribbean as a region and Jamaica in particular, are significantly different from those in the US and Europe.

The Harder They Come tells the story of Ivan, played by Jimmy Cliff, who comes from rural Jamaica to Kingston to become a musician. In fact, Jimmy Cliff's story is similar to the fictional Ivan's, in that neither man was paid for his first recording. In the film, Ivan receives a contract from Hilton that would give him only $20.00 for all rights to the recording and he refuses to sign it. But when he tries to sell his record himself, he finds out that Hilton controls all the record stores and radio stations in Kingston, so he ultimately returns to sign a contract that gives Hilton only the performance rights, but still only for $20.00. When his talent and ambition are exploited, he gets involved with the drug trade. When he realizes that the drug trade is also exploitative, and so opposes his boss in the drug trade, the corrupt police force tries to kill him, so he flees, hides, and becomes an outlaw. The more lawless he becomes, the more the masses celebrate his persona. Subsequently, the recording he made, but which was held back by his producer, becomes a hit. Those who control the drug trade, however, try to starve out the small-time drug-traffickers so that they go hungry and give up Ivan. In the film's dramatic ending Ivan tries to flee to Cuba but tragically cannot reach the ship, is found by the police and killed.

The film has cult status primarily for its soundtrack and the participation of the Reggae artists Toots and the Maytals, Desmond Dekker, and Jimmy Cliff. Its fame is achieved in part through the music, which is integral to and is integrated into the narrative. *The Harder They Come* juxtaposes the music made by those whose only resources are their own talents and voices with the record industry that owns the means of cultural production but exploits the local bands and singers. When Ivan attempts to make a reggae record, we first see him at the gate of the recording studio, where Hilton, the only music producer on the island, makes poor black men sing to see whether he is interested in recording their music. Ivan gets his break when he records the song (written, like the soundtrack, by Cliff) which gives the film its title and provides its musical motif and refrain:

> Well, you tell me about pie up in the sky
> Waiting for me when I die
> But between the day you're born and when you die
> Never seem to hear even your cry
> So as sure as the sun will shine
> I'm gonna get my share, what's mine
> And then the harder they come
> The harder they'll fall one and all
> Well, the oppressors are trying to keep me down
> Making me feel like a clown
> And they think that they've got me on the run
> I say "Forgive them, Lord they know not what they've done"
> Cause as sure as the sun will shine
> The harder they'll fall one and all
> And they think that they've got me on the run
> Tell them that they don't it ain't no fun
> And I'd rather die than live and be a slave
> Yes, and I'd rather be right in my grave . . .

The film tells the story of the power of the individual to resist and advance his own vision with his personal abilities as his only resource. *The Harder They Come* relies on the performance of one person to embody with his voice the singular vision of a resistance to the system of exploitation. Ivan is a young man with a dream, and the film shows how such urban dreams are fed by the media, in this case a film within the film and the transistor radio, but also how Ivan makes use of the media to advance his dream in the second half of the film. The film within the film shows a Spaghetti Western, and the director, Perry Henzell, explains: "Spaghetti Western was an obsession in Jamaica for years, before Kung Fu. Spaghetti Western was the staple of slum cinema, the same formula where the hero doesn't really want to fight but they push him and push him and push him until he starts to fight" (DVD commentary). At the level of the narrative, Ivan's visit to the movie theater in

continued

Kingston, "The Rialto," is his first encounter with the possibilities of the city. But on a meta-level the consumption of the Spaghetti Western in Jamaica also shows the transnational exchange of cinematic images in the margin of society but also the lack of local figures of identification for the local, poor, Black, and Jamaican audience. The scene at the Rialto cross-cuts clips from the Spaghetti Western with the reactions of the audience showing the consumption of popular cinema in terms of identification. *The Harder They Come* shows the active identification of the all-Black audience with the white heroes of the film within the film and reflects on the conditions of film consumption, race, and national identity and culture in Jamaica, where no films with or by Jamaicans in general, or Black Jamaicans, or poor Black Jamaicans for that matter, were available before *The Harder They Come*. The film's opening in Kingston was an important moment because "Black people seeing themselves on the screen for the first time" created an "unbelievable audience reaction," Henzell points out on his DVD commentary.

The Spaghetti Western fuels Ivan's fantasy, including in the final shoot-out in which he is killed. That shoot-out is intercut with shots of the audience watching the Spaghetti Western from the beginning of the film, showing that Ivan has reached his dream of becoming a legend, and that he now has to pay the price, death. The outlaw wants what he cannot get, justice and equal distribution of resources, and when he understands that he cannot get it he willingly dies. The killing of Ivan is not fetishized: Henzell avoided the use of slow motion but instead took out shots so that Ivan's death occurs more quickly. The film then immediately cuts to a woman dancing to the music of Jimmy Cliff/Ivan and transfers the diegetic legend of Ivan to the music that did become legendary.

The city of Kingston, Jamaica, is marked by class divisions and the simultaneity of local culture and global influence. This is encapsulated in the early shots of the city, when Ivan arrives by bus from the countryside. On the one hand, we see billboards for Philips and Shell that announce the modern urban space with global industrialized products. On the other hand, the young man who steals Ivan's belongings immediately after his arrival has a pushcart with which he moves through the city, on foot, and which he has adorned like a car. Henzell explains that this represents "a defiance of poverty," because "you can be poor and you can be proud" (DVD commentary).

The Harder They Come confronts economic and cultural injustice and opens up a fantasy space, but is indebted to the politics of realism associated with traditions of political filmmaking. Most of the film is shot in the real ghetto of Kingston, which consists of makeshift shacks, alleys, and waterways lined with trash. Lives are lived in intermittent spaces between inside and outside. Ivan straightaway meets people who play dominoes outside but who cannot pay their debts. When something happens, people gather immediately. Gates signify the separation between the ghetto, with its inhabitants, and other neighborhoods, as when the singers wait at Mr Hilton's gate, or when Ivan enters the garden in a high-class neighborhood to offer his services and after he has left, the mistress yells at her servants for having left the gate open. When at first Ivan cannot get work, he lives with the preacher to whom he was sent by his mother, but his favorite place is an abandoned car, which he transforms into a space of his own with a toy gun, comics, and a *Playboy* magazine. His prime possession is a bike that he repairs and with which he takes Elsa, the daughter of the preacher, to the beach. The preacher has looked after Elsa, an orphan, and when he finds out that she and Ivan are in love, he kicks Ivan out and gives his bike to Longa. Ivan and Longa fight over the bike, which leads to Ivan's arrest and public whipping, which conjures up images of slavery and emphasizes his humiliated Black masculinity just before the film's turning point, when he sings his hit song in the studio. *The Harder They Come* portrays the limited choices that ultimately lead Ivan to become involved in the drug trade without celebrating illegality.

Figure 7.3 Perry Henzell. *The Harder They Come* (1972): Running through shacks

continued

Most of the footage of the ghetto, the market, the dump, Kingston in the rain, consists of images that are not staged or composed for the camera. Such on-location shooting makes a political statement, as pointed out by Keith Q. Warner in *On Location: Cinema and Film in the Anglophone Caribbean*, since the Caribbean has traditionally been used as the exotic backdrop for films without representing the reality of life in the Caribbean. In addition, however, on-location shooting also reflects the reality of low budgets and directors who cannot afford settings built in the studio. One striking indication of the film's low budget is the use of moving toy cars on a map of Kingston to show the police chase of Ivan. What would seem absurd in other films seems at that point in the narrative entirely normal to an audience engrossed in the story and used to the improvisation in low-budget independent cinema.

The Harder They Come empowers the self-representation of poor, Black Jamaicans. Once Ivan has been chased by the police, he goes to a photographer and has his picture taken dressed as an outlaw. At the studio he meets a man who wants his autograph and he sends the picture to the newspaper endorsing a fantasy self-representation in a film characterized by the reality of poverty. Since the people now protect him, and his song plays on the radio, there are signs everywhere saying "I was here but I disappear." The film evokes the Jamaican mythology of Rangyn, Johnny Too Bad, and thus functions within a collective, continuing a legend. Towards the end, Ivan steals a car and drives around a golf course, acting out his fantasy of owning a convertible. We see him driving in circles on the green lawn, with the music playing, clearly acting out a fantasy not just for himself but for the audience as well. The audience understands that Ivan's self-fashioning will have a tragic end. He becomes a victim but a victim with agency, a will, creativity, choices, and a clear voice that is made legendary through the film.

Further reading

"Blaxploitation Revisited" (2005) *Screening Noir: Journal of Black Film, Television, and New Media Culture* 1, 1 (fall/winter). A collection of current theoretical approaches to Blaxploitation.

Gerald Martinez, Diana Martinez, and Andres Chavez (eds) (1998) *What It Is . . . What It Was! The Black Film Explosion of the '70s in Words and Pictures*, New York: Miramax–Hyperion. This is not an academic book but a collection of visual materials for fans, which can be useful to those who have no visual archive of Blaxploitation.

Douglas S. Massey and Nancy A. Denton (eds) (1993) *American Apartheid: Segregation and the Making of the Underclass*, Cambridge, MA: Harvard University Press. Massey's and Denton's volume coined the term "American Apartheid" and provides the background to the re-emergence of the urban ghetto in the 1990s.

Paula J. Massod (2003) *Black City Cinema: African-American Urban Experiences in Film*, Philadelphia, PA: Temple University Press. Massod's volume provides an extensive overview of African-American urban cinema that begins with films from the early twentieth century, addresses Spike Lee's cinema, and moves up to the late 1990s. It fills in some of the gaps of this chapter.

Keith Q. Warner (2000) *On Location: Cinema and Film in the Anglophone Caribbean*, London: Macmillan Education. Warner situates *The Harder They Come* in the context of the Anglophone Caribbean film industry. He emphasizes the contrast between the representation of the Caribbean as a colorful backdrop to European and American cinema and the self-representation of Caribbean filmmakers.

Essential viewing

Charles Burnett. *Killer of Sheep* (1977)

Haile Gerima. *Bush Mama* (1976)

Perry Henzell. *The Harder They Come* (1972)

Fernando Meirelles. *City of God* (2002)

John Singleton. *Boyz N the Hood* (1991)

from B. Mennel, Cities and Cinema (Routledge, 2008)

8 The city as queer playground

Lets go to bed with each other! Or fool around in parks if there are no beds. Boys with girls, boys with boys, girls with girls, men with boys and girls, women with men or boys or girls or tamed little panthers – what's the difference? Let's embrace each other! Let's dance!

Klaus Mann

The city is one of the crucial factors in the social production of (sexed) corporeality: the built environment provides the context and coordinates for contemporary forms of the body.

Elizabeth Grosz

Learning objectives

- To understand the changing relationship of sexuality and the city
- To conceive of sexuality as intersecting with race, class, and gender mapped onto the city
- To distinguish postwar shifts in the cinematic representation of gays and lesbians and queer desire

Introduction

This chapter's title invokes the playful and liberating aspects associated with the city in contemporary gay and lesbian culture. In 1969 the Stonewall riots marked the earlier historical turning point from homosexuality as a secret, underground activity to "out" lesbian and gay identities. "Stonewall" refers to the urban rebellion of drag queens on the piers of New York City against the police that enabled gay and lesbian life to be "out," the public display of gay and lesbian identities, affections, and lifestyles. In 1981, the AIDS crisis gave rise to radical queer activism, which enabled the queer culture of the 1980s and 1990s, including an explosion of queer cinema. AIDS was also the context for films that presented a dystopian view of a dark underworld defined by sexual perversion, alienation, violence, discrimination, and deviance, mobilizing homophobic fantasies of an urban gay and lesbian underworld that can kill on contact or turn innocent characters into killers, for example in Brian De Palma's *Dressed to Kill* (1980) and William Friedkin's *Cruising* (1980). Gay and lesbian directors have since developed a diverse cinema that circulates transnationally and ranges from low-budget, avant-garde, political films, to mainstream genre films. This chapter discusses representative examples of films that mark these shifts regarding the ways in which sexual desire and non-normative sexual practices are cinematically mapped onto the city.

The sexual city

From the inception of modernity, cities, and particularly their public spaces, for example the streets, have been sexualized, as this volume has outlined. But certain world cities have always been particularly associated with gays and lesbians: Amsterdam, London, Paris, Berlin, Alexandria and Cairo, Istanbul, Havana, Rio de Janeiro, San Francisco, New York City, and Toronto are the best-known examples. In some instances, it is not the entire city but a particular neighborhood that is associated with the consumption of sex through porno theatres or with gay and lesbian lifestyles, such as gay bars. Most of the oeuvre of gay enfant terrible John Waters celebrates the seedy culture of Baltimore's Fells Point, a former harbor neighborhood that was originally frequented by sailors and catered to them with B-film movie-houses and bars. John Rennie Short points out that cities include areas in which "sex is a commodity and alternative sexual practices take place," for example the "gay ghetto," which is sometimes associated with gentrification (136). Identity categories of race, class, and sexuality intersect in the shaping of neighborhoods but they can also create divisions. Linda Goode Bryant and Laura Poitras's film *Flag Wars* (2003) documents conflicts between gay white men who are at the forefront of the gentrification of a poor, Black neighborhood in Columbus, Ohio, and the original inhabitants of the neighborhood who are being displaced.

Short explains that it is difficult to find hard data about the sex industry, which is often located in cities (137). There is a close connection between the porn-film industry and B-film movie-houses in cities, prior to the video and internet revolution, both of which led audiences increasingly to consume sex films in the privacy of their homes and decreased the need for porn movie-houses in cities. Prior to these changes certain parts of cities were associated with B-film and porno movie-houses, in proximity to strip clubs, night clubs, prostitution, and brothels. Films shown in those movie-houses included Blaxploitation, Hong Kong action films, cheap science fiction, soft-core porn, gay, lesbian, and straight sexploitation, horror, and trash cinema. Among porn, scholars differentiate between hardcore, which shows heterosexual penetration, and different kinds of soft porn or sexploitation genres. In the 1960s, a subgenre ostensibly re-enacted educational narratives, such as Doris Wishman's *Bad Girls Go To Hell* (1965) or Joseph P. Mawra's *Chained Girls* (1965). These films pretended to educate viewers about deviant behavior as a pretext to show titillating images. The tradition of integrating narrative, documentary footage, and educational information can be seen as early as 1919 in Germany's gay silent film *Different from the Others* (Richard Oswald, 1919), which mixes footage from a gay bar in Berlin in the 1910s and a lecture by Magnus Hirschfeld, sexologist and gay rights activist in Weimar Germany.

Different from the Others is set in Berlin, where the main character, Paul Körner, a music teacher, falls in love with his student, Kurt. Their visit to a gay bar is presented through documentary footage from a gay bar in Berlin in the 1910s that shows same-sex couples in costume dancing. After their visit, Körner is blackmailed because homosexuality is outlawed. The film paradigmatically integrates a story with documentary footage, and activism with educational enlightenment. It addresses the gay and lesbian community by offering a positive self-image on the screen. It also aims at the wider public, urging educators, liberals, and politicians to reconsider Paragraph 175, which made male homosexuality illegal in Weimar Berlin. The film offers a glimpse of the extensive and famous Berlin gay-bar scene of the time. But the bar is also the place where the blackmailer spots Paul Körner. From the inception of gay film the gay bar has been coded as a double-edged collective pleasure and danger.

Short emphasizes the "tension between emancipation and commodification" of sexual culture in the city, which also applies to gay and lesbian film culture (139). Many young gays and lesbians nowadays experience gay and lesbian normalcy, surrounded by queer characters in mainstream film and television. They are less dependent on the anonymity of the city and its sexual subculture. On the flipside of these developments in the last couple of decades, particularly in the West, gay and lesbian culture has been increasingly commodified. The emphasis of this chapter however lies with lesser known, independent cinema. Social developments influenced the representation of gay and lesbian identity and non-normative sexual desire on the screen: pre-Stonewall, queer desire was a cinematic subtext to be encoded by directors and decoded by a subcultural gay and lesbian audience. Gay and lesbian audiences were often forced to read against the grain imposed by production codes that led to homophobic and stereotypical depictions of the gay and lesbian community. Today we find overt representation of gay and lesbian characters and narratives. Even though gays and lesbians are less dependent on the city in contemporary times, much of the media representation continues the tradition of linking sexuality to urbanity by placing depictions of gay and lesbian communities within urban contexts.

1960s–1970s: homophobia and gay desire in the city

As outlined in this book's Introduction and Chapter 1, from its inception the cinematic city represented pleasure and danger. This double characteristic is especially pronounced in the portrayal of the city and its liberating, but also dangerous, even murderous, sexual subcultures. Gay and lesbian cinema has undergone radical changes throughout the twentieth century from its beginnings in Weimar Berlin to the repressive films of the 1950s in the USA, the films that represent gay and lesbian liberation in the 1960s, to the explosion of queer cinema in the 1980s and 1990s, and finally to the commodified television shows and mainstream gay and lesbian characters that populate mainstream film now.

Prior to the queer cinema of the late 1980s and early 1990s, much of mainstream depictions of gay life in the city were deeply shaped by homophobic anxieties. The danger of the bar is at the center of such films as *Cruising* (1980), which refers to the gay male practice of looking for men for anonymous sex in the city, in either bars or public parks. *Cruising* uses the gay male subculture of New York City as the setting for an action narrative in which a detective has to go underground to catch a man who engages in sadomasochistic practices with and then kills his sexual partners. Here the city is synonymous with gay anonymous sex, which in turn is associated with murder. The film is part of a one-dimensional homophobic discourse that equates gay male sexuality with death, a dangerous equation that mirrored public anxieties during the emergence of the AIDS epidemic.

John Schlesinger's earlier *Midnight Cowboy* (1969) paints a more sympathetic picture of its characters and the city, and complicates the relationship between the rural and the urban. Joe, an innocent and naïve young man from Texas takes a bus to New York City, where he intends to work as a – heterosexual – hustler. When his plan fails, the homeless Rizzo helps him survive in the city by showing him how to "squat" in empty buildings. Rizzo's health is failing, and in order to save him Joe finally turns a gay trick (has sex for money) but kills his client for his money, so that he and Rizzo can take a bus to Florida. On the way Rizzo dies.

This film makes recourse to the stereotypes of the wholesome country, here embodied by Joe, a blond, attractive, naïve Texan from a small town, and the decadent and anonymous city, embodied by Rizzo, a short, dark, run-down, disabled Italian. At the same time, *Midnight Cowboy* deconstructs these stereotypes by making the city the space in which their friendship grows and the country the origin of the traumatic, sometimes sexual, nightmares that haunt Joe. The film's opening shot creates rural America as an expansive space with a humongous billboard. *Midnight Cowboy* reverses the traditional Western narrative, as the journey takes our main character not from the east to the west but from the west to the east. As in *The Harder They Come*, we see the rural to urban migration happen when the caretaker, in both cases the grandmother, dies, and as in *The Harder They Come*, a small transistor radio is the lead character's fetishized property. Here it provides the audio background for the changing social spaces from the rural west to the urban east of the USA. Different stations provide the cultural soundtrack for the transition and journey across the country. After Joe has arrived in New York City, he listens to a radio broadcast of a call-in show of women discussing the men they like, which suggests that the women of New York know, articulate, and actively pursue their sexual desires. When Joe shares his life with Rizzo in New York City and they have to pawn their belongings, the transistor radio is the last thing to be pawned.

In *Midnight Cowboy* the difference between the country and the city is demarcated not so much by the presence or absence of sex, but by the ways in which sex is negotiated. The film suggests that in the city sex is open and liberated, while in the country it is repressed by individual and systematic violence. The city is inhabited by emancipated and sometimes unscrupulous women who take advantage of Joe. The first woman whom Joe approaches tells him that he should be ashamed of himself. The second woman takes him to her penthouse apartment, where she talks on the phone with a man while Joe undresses her. While they have sex on the bed, the camera pans to the television, which shows a series of shots surfing through the different channels, culminating in a gambling machine that spouts money, a parody and reference to the "money shot," the shot in pornography that explicitly shows genitalia in the moment of orgasm. The shot creates an ironic intertext for those who know about the "money shot" by referencing pornographic conventions in a mainstream film to comment on the fact that Joe is selling his body and performs sex for money. Yet, after sex, the unnamed woman sheds tears and manipulates Joe to give *her* money for a cab to see her next date. Here the film relies on the stereotypical dichotomy of the urbanite who can exploit the naïve character from the country.

Joe's past in Texas is hinted at in gothic flashbacks that cannot be pieced together entirely and which in their fragmented form reflect traumatic events: a girlfriend who loved him but who was gang-raped while he was held down and possibly also sexually assaulted; a mother who left him behind with his grandmother; and that

Figure 8.1 John Schlesinger. *Midnight Cowboy* (1969): Dustin Hoffmann ("Ratso" Rizzo) and Jon Voight (Joe Buck) on the set

grandmother who left him home alone while she went out with her sexual partners, kept him in bed with her men, and performed medical procedures on him. These flashbacks hint at childhood sexual abuse that led to Joe's self-image as someone who was only good in the arena of sex. *Midnight Cowboy* plays with the idea of the city as sexual playground but depicts this idea as Joe's misconception. It is his rural naïveté that makes him a victim of sexual exploitation, particularly by the women around him.

Walking the streets of the city structures *Midnight Cowboy*, a motif that has reoccurred throughout this book. The film shows Joe alone, and later with Rizzo walking through the city, without dialogue or particular narrative development, one element that the following films share: *The Last Laugh*, Ángel Muñiz's *Nueba Yol* (1995), *The Harder They Come*, *Lola and Billy the Kid*, *Boyz N the Hood*, *The Horse*, *Cléo from 5 to 7*, *Brothers and Sisters*, *Bush Mama*, *Shaft*, and R'anan Alexandrowicz's *James' Journey to Jerusalem* (2003) (discussed in Chapter 9). With the exception of Cléo, who is privileged and who finds her true identity through the city, all of the characters of these films are disenfranchised. Their walking is often necessitated by the lack of home and shelter, by an insufficient home as in *The Harder They Come*, or by lack of money for transportation as in *The Horse*. This kind of walking through the city is different from that assiociated with the concept of the *flâneur* suggested by Walter Benjamin, where the subject has a sense of being pleasantly lost in the city, lured by its attractions, and where desire is embodied by the prostitute, who is always assumed to be female. This is sometimes invoked when Joe walks along the old 42nd Street passing the gay, male hustlers, but the overarching reason for his movement through the urban public space is want and disenfranchisement. These films capture the ambivalence of exclusion, the lack of resources, the attraction of the spectacle of the city, and participation in anonymous crowds. Because Joe himself is a prostitute, his walking is driven by despair, not by seduction.

In *Midnight Cowboy* the intersection of commodity and sexual outcast is captured when black-leather-jacketed, young artists take a photo of Joe and Rizzo at a diner, seeing them as "real" urban down-and-outs. Joe and Rizzo are then invited by the artists to a party where guests do drugs and photos are projected onto the wall, and where the hosts film the guests, including Joe and Rizzo. Joe is picked up by another self-confident woman while Rizzo, whose health is deteriorating, drags himself to their squatters' apartment. Joe cannot sexually perform for the woman who finally uses him as a hustler, which reflects his growing marginalization and emasculation, and the film's hinted but not clarified gay desire articulated as care for and intimacy with Rizzo.

In order to help Rizzo to get to Florida, Joe finally works as a gay hustler. While his fantasy of being a straight hustler was accompanied by his notion of sexual prowess and successful masculinity, his work as a gay hustler is portrayed as a sacrifice for Rizzo and the low point of Joe's life in New York City. The specter of the gay hustler has been announced throughout the film, and Joe's homosexual hustling emasculates and marginalizes him, and ultimately turns him into a killer of his older, repressed, gay client in a cheap hotel room in the area of New York City's 42nd Street, coded as a violent and dangerous space. While the film promises to rewrite of the urban–rural binary at the outset, it ultimately vilifies the city as a site for homosexual, anonymous sex and homophobic violence, with which we are, however, to identify via Joe. Paradoxically, the relationship between Rizzo and Joe is portrayed as one of care, empathy, concern, and understanding in a desexualized manner that nevertheless remains ambivalent about their attachment to each other. Many of the other characters refer to Rizzo and Joe as gay, but Rizzo consistently makes homophobic remarks to distance himself from "faggots." One must infer a kind of ambivalent gay desire that is simultaneously homophobic. This tension is narratively resolved through the violent act against the character onto whom a coherent gay identity is projected, the repressed homosexual who operates in the anonymity of the city, who cannot engage in love, and instead buys sex with money. The film vilifies the kind of anonymous and commodified sexual relationships that the city enables.

1980s–1990s: new queer cinema and transnational urbanity

In contrast to the films from the late 1960s and 1970s that tie gay desire to homophobic discourse, the new queer cinema that emerged in the later 1980s and 1990s celebrates a queer life that validates the heterogeneous possibilities of the city in a transnational perspective, for example Monika Treut's *Virgin Machine* (1988) and *My Father Is Coming* (1991). This new queer cinema also reveals those gay and lesbian aspects of urban history that have been denied or forgotten, as in Isaac Julien's *Looking for Langston* (1988). Both films by German lesbian filmmaker Monika Treut, *Virgin Machine* and *My Father Is Coming*, show the main character migrating from Germany to the gay and lesbian urban centers of the US, San Francisco and New York City, where she experiences liberated and self-confident sexual expression. In *Virgin Machine* the main character, journalist Dorothee Müller, travels from Hamburg to San Francisco in order to research "romantic love." Instead of a realistic narrative, the plot meditates on the changing notion of love in late twentieth-century bourgeois society. The dividing line between romantic love and sexual play, however, coincides with the national boundaries of Germany and the USA.

The film begins in Hamburg with Dorothee in a relationship with an overweight, unattractive, unappealing, sweaty man, shot only from camera angles that distort him, and then she is involved with Bruno, her half-brother, who we also see in a gay encounter in a public bathroom. She then travels to San Francisco to research romantic love and find her mother. Her mother has disappeared, however, and while she is staying in a hotel, she sees an advertisement on television for "Ramona," which suggests, "You could be addicted to romantic love. Call now." Dorothee calls Ramona with the explanation that she is from Germany writing an investigative report. The two spend a romantic day on the pier together and finally have sex. When Ramona leaves, she gives Dorothee a bill – for her company, entertainment, and sex. Dorothee also visits a strip club in which women perform for each other, dressed up as men or as women, and Susie Sexpert shows her different kinds of sex toys. After watching a striptease for women, women cross-dressing as men, and sado-masochistic rituals in other hotel rooms, and making friends, Dorothee ends up at the Golden Gate Bridge, where she tears her old photos into small pieces.

San Francisco developed into the national and international center for gay life because of its strategic position as a harbor and military base which attracted both men and women who did not want to live by society's gendered norms in the rural USA, and it became a city where gays and lesbians could enjoy rights equal to those of heterosexual couples and thus attracted more gays and lesbians from around the world. *Virgin Machine* captures the moment in San Francisco in which the notion of a gay, male life that had dominated the 1960s and 1970s and was decimated by AIDS in the 1970s and 1980s gave way to a queer culture that brought together feminism and lesbianism and other kinds of marginalized sexual practices in the 1980s and 1990s. The tragic deaths of so many gay men in the 1980s and the ensuing activism, which included lesbians, led to the reappropriation of the term "queer," previously a discriminatory and derogatory term that now includes a range of non-normative behaviors and desires, moving beyond gays and lesbians to include bisexuals, and transgender, transsexual, and cross-dressing subjects.

During the 1960s and 1970s, second-wave feminism had regarded sexuality as one of the areas in which patriarchal norms reproduced themselves through the violation and objectification of women. With regard to the topic of cinema and the city, this attitude can best be illustrated by the resulting public protests at porn movie-houses, when groups of feminists aligned themselves with right-wing politicians and family-value anti-pornography activists to denounce the demeaning effect of pornography on women. This attitude changed radically in the late 1980s, when feminists embraced sexuality and, as with the term "queer," appropriated the position of the voyeur by claiming the pleasure for lesbians to look at sexualized women and the pleasure for straight women to look at sexualized men. These third-wave feminists also claimed the role of the sex objects as a potentially pleasurable and empowering position, and appropriated masculinity through cross-dressing, and finally claimed also the pleasure in pornography by creating feminist straight and lesbian pornography.

Virgin Machine gives us one of the few representations of a lesbian hustler on film. Films, such as *Midnight Cowboy*, show aggression as the response to the confusion around the sex that is for sale but is mistaken for love or desire. In *Virgin Machine*, when Ramona presents her with the bill after a day of pleasure, Dorothee breaks out laughing. Her laughter captures the film's attitude towards the commodification of sexuality and desire between women that is enabled by the metropolitan city: not disappointment, nor aggression, nor a melancholic sense of loss, but rather irreverence. The theorists discussed in this book have been critical of the commodification of sexuality, but at the same time have fetishized the figure of the prostitute as the prime example of that commmodification (compare, for example, Benjamin). Instead of criticizing the commodification of women's sexuality as the feminists of the second wave of feminism did, the queer sex-positive feminists like Monika Treut reappropriate the figure of the prostitute and portray her as a self-confident negotiator and enabler of desire.

A similar strategy underlies the encounter between Dorothee and Susie Sexpert, a performance artist and author intent on liberating women's attitudes towards sexuality. The camera shows two men watching Dorothee and Susie Sexpert's discussion about sex toys in the street, with neither voyeuristic delight nor disgust, but with a quiet curiosity. Taking the discussion of the pleasure of sex into the street involves several transgressions for women: taking sex from the private to the public sphere, portraying women as self-confident participants in such conversations, and presenting sex toys as clean, feminist, and driven by fantasy, in contrast to the discourse about sex forced into the dark urban zones of overpriced, dirty, unhealthy, nightly activities by sexually repressed bourgeois men as portrayed in Joe's customer in *Midnight Cowboy*.

Virgin Machine deconstructs the idea of romantic love via queer pleasure found in the city of San Francisco, which becomes the site of a queer *Bildungsroman* for Dorothee. Her voice-over states her political manifesto: "Susie Sexpert is right. The sex industry is so lousy because women have no say. Feminists should go there instead of being uptight. It's the perfect place to live out their fantasies." The city of San Francisco becomes a utopian space for a feminist and queer agenda, and documentary *cinema verité* strategies are used to claim the authenticity of the historical and geographical moment of San Francisco in the late 1980s: the city had long represented a sort of sexual utopia, but this was a radical transformation in San Francisco in the 1980s for women. When at the end of the film Dorothee tears up photos and throws the scraps into the Bay under the Golden Gate Bridge, she is tearing up a private visual representation of desire in order to meld with the queer community. She has become a transnational queer who does not need to be attached to a home or a romantic past. Her character embraces the commodified exchange of desire and sexuality, traditionally denounced as the dirty underbelly of urban life, as a nomadic, postmodern sexuality. The film offers a fantasy of liberated sex and gender with neither violence nor deprivation, presenting the city as queer playground.

The city as queer playground also includes the possibility of alternative family structures. Takehiro Nakajima's *Okoge* (1992) – the title is the Japanese term for straight women who are attached to gay men – offers a complex portrayal of straight female and gay male desire, of femininity and masculinity, of liberation and the traditional values of family and work. It begins at a beach where gay men are enjoying themselves. Sayoko is a young woman who is intrigued when she sees gay men kissing and befriends the couple, Goh, the only son of a widow, and Tochi, an older, married man. When Goh's widowed mother moves in with him, Goh and Tochi have to find another place to meet secretly, and Sayoko offers them a room upstairs in her little house. Goh and Tochi make love upstairs while Sayoko sits in her bed downstairs and looks at pictures of Frida Kahlo in an art book. We are shown haunting memories of her childhood when she was abused by her father, offered as an implied explanation for her triangulated relationship with the two gay men. Goh and Tochi break up when Tochi's wife threatens to expose his homosexuality at work, and Goh falls in love with a straight man who rapes Sayoko and then leaves her with a child. Sayoko disappears from Goh's life until he hears about her financial and emotional difficulties as a single mother.

Walking through the rainy city with his cross-dressing friends in full drag, Goh runs into Sayoko as she is being harassed by debt-collectors. Goh fights with them, and then the transvestites beat them up with their handbags and high heels. In contrast to films in which we find either a homophobic portrayal of a murderous cross-dresser (*Dressed to Kill*) or homophobic violence against cross-dressers (see case study of *Paris Is Burning*), the image of the angry transvestites is empowering to viewers who identify with them. They are happy about their victory but mourn their ruined make-up, and Goh invites Sayoko and her baby to live with him and offers to act as father to the baby. The following scene shows a wedding with Tochi and a cross-dressing transvestite as his wife in front of his colleagues at work, and the final scene of the film shows Goh and Sayoko walking through a street of the gay neighborhood in Tokyo with the baby between them. We are positioned in the midst of the crowd with men kissing, interracial couples walking hand-in-hand, and cross-dressing men in traditional kimonos being affectionate with each other. Goh and Sayoko embody an alternative family that is integrated within the eroticized gay urban space. The film's ending thus offers a utopian vision beyond the binary of hetrosexuality projected into a private and sexualized homosexuality projected into a public space.

Most of the films considered in this chapter show us maturing or mature characters who move or travel to the metropolis to develop a gay, lesbian, or transvestite identity in a community that supports the individual in alternative social structures. New York, Tokyo, Buenos Aires, Berlin, and San Francisco become the setting for anonymous exchanges of desires, non-normative identity formation, and urban communities. The increasing social acceptance of gays and lesbians throughout the 1980s and 1990s makes queers less dependent on the anonymity of the city for their self-expression. Thus, films addressing gays and lesbians can now situate their stories in other spaces, suburbia, for example. Alain Berliner's film *Ma vie en rose* (1997) is set in a Belgian suburb and tells the story of Ludovic, a boy who thinks he is a girl. The film suggests the possibility of retaining one's birth family, with its care and attachment in the face of non-normative desires of children that threaten the normative social order, played out not in the city but in suburbia. Ludovic is 7 and lives with his family, which consists of his parents, siblings, and grandmother. The film is shot in the bright colors of afternoon children's television shows and integrates a doll, Pam, who lives in a Barbie-like plastic house. Ludovic innocently believes that he is a girl and wants to marry a boy. The family's earnest attempts to understand give way to a crisis when Ludovic's father is fired by his boss who is unwilling to accept a family with a male child who wants to wear make-up and skirts.

The film portrays the subtle violence of well-meaning, middle-aged liberals in modern suburbia who have organized their desires in well-functioning nuclear families. Ludovic's parents obviously love him and are alternately kind and supportive, and aggressive and prescriptive. They are forced out of the upper-middle-class neighborhood with clean houses and perfectly manicured lawns into a neighborhood of a lower socio-economic standard. There at a party Ludovic meets a little girl who likes to dress as a boy and he exchanges his cowboy costume for her princess dress, and the new neighbors cannot understand why the family has been so upset when Ludovic cross-dressed. The film makes a case for more subtle negotiation of homophobia and gender-normative behavior, and moves away from the depiction of clear-cut violence, focusing on the normative politics of class and the hypocrisy of enforced gender norms. While *Ma vie en rose* acknowledges the psychological labor that goes into working through homophobia, it ultimately creates a hopeful vision about the possibility of family as a source of support, unlike such films as *Virgin Machine*, which leave notions of family and romantic attachments behind. In *Ma vie en rose* family is possible, though not easy, when its members care about each other's well-being more than about material gain and living in the "right" neighborhood.

The city in the present promises sexual liberation but a gay past is filmically invisible, and for ethnic, migrant, and racial subjects that liberation is a more complex process than is suggested by Treut in *Virgin Machine*. Isaac Julien was part of a group of emerging filmmakers of color in Britain, some of whom were queer, who were funded by Channel 4 in the late 1980s and early 1990s for their aesthetically and politically innovative films. His *Looking for Langston* (1988) is a transnational project with Black, gay masculinity at its center, so it is perforce set in Harlem and the historical point of reference is the Harlem Renaissance. While the film pinpoints a crucial time and place for the articulation of Black, gay desire, it simultaneously points to the limits of representation by integrating historical footage with fantasy scenes of gay, Black life re-enacting the past. *Looking for Langston* is subtitled "A Meditation on Langston Hughes (1902–1967) and the Harlem Renaissance//With the poetry of Essex Hemphill and Bruce Nugent (1906–1987//In Memory of James Baldwin (1924–1987)," and captures the process of remembering through emotional attachment. The film weaves a tapestry of voices and re-enactments, and thus rereads the cultural production – particularly the poetry but also the urban space – of Harlem, which is conventionally marked solely as Black, and therefore straight, in the public imaginary: "Homosexuality was the sin against the race, so it had to be kept a secret, even if it was a widely shared one."

Looking for Langston begins and ends with documentary footage of the funeral of Langston Hughes, framing the film in part as a eulogy. By restaging the funeral and creating an imaginary space that includes an upstairs with the corpse and a downstairs with a bar, Julien creates a metaphor for urban gay life that is celebrated as long as it does not act out its desire publicly. The film thus presents an image of the Harlem Renaissance that expands the general understanding of it solely as cultural expression of Blackness. It also reinterprets the notion of urban space by extending the urban gay brotherhood with a fantasy world of rural beauty, where naked men encounter each other in fields of poppies. These images alternate with shots of urban nights when individual men walk through parks and the empty streets of warehouses, the sites of anonymous encounters. The voice-over quotes Essex Hemphill's poem *Where Seed Falls*: "In the dark, we don't have to say 'I love you'" (Hemphill 159). The poems by the late poet Hemphill and by Harlem Renaissance poet Hughes, read over contemporary images, bring together Black gay culture from the early and the late twentieth century. But like other films, *Looking for Langston* references homophobic violence: "But love is a dangerous word in this small town, those who find it, are found face down," again quoting Hemphill.

While there are several shots of Harlem's main streets in the film, the general setting is an interior fantasy space coded as beyond time. Beautiful Black men dressed in tuxedos are dancing with each other, and as the film advances they move into modern dance movements to contemporary music. The dangerous urban space that represses homosexual desire is contrasted with the liberated self-expression and celebration of beauty possible in interior spaces. The expression of Black, gay, male desire contradicts the notions that gayness is white and Blackness is straight. "Touch me now, I am a revolution without bloodshed, I can be an angel falling," explains the voice-over, quoting Hemphill. But then the tone changes and the dancing suddenly stops, and outside we see a group of men in contemporary neo-Nazi outfits, accompanied by British police, breaking into the bar and threatening everyone. The film returns to the opening of Langston Hughes's funeral, again contrasting and putting in dialogue the past and the present, the urban exterior and interior. The achievement of Julien's film, in the context of the questions this book raises, lies with the reinterpretation of an urban space and time period that traditionally had been seen solely as Black. Julien represents Harlem and the Harlem Renaissance as Black and gay and in that process changes the meaning of the terms Black, gay, Harlem, and Harlem Renaissance.

Turkish filmmaker Kutluğ Ataman's *Lola and Billy the Kid* (1999) is similarly concerned with the urban landscape, but that of Berlin, a city traditionally associated with gay liberation, as portrayed through the lens of a minority group, in this case of Turkish-Germans. Like many of the films discussed in this chapter, *Lola and Billy the Kid* tells the story of a group of young, gay, migrant, Turkish-German men, the grown-up children of "guest workers" in Germany. Several of the men work as hustlers in gay bathrooms at the Hermannplatz in Neukölln, a predominantly working-class neighborhood of former West Berlin. The other characters form a cross-dressing cabaret group that performs exaggerated orientalized femininity. They all move about the city of Berlin through spaces that are coded as gay sex zones in the gay subculture, such as the Tiergarten, a big park in the west. The film begins with a shot of the victory column (Siegessäule), which is also the name of the gay magazine of the city, but most of the intimate scenes are set in interior spaces. Lola, a major character who cross-dresses as a woman, returns to Kreuzberg, the neighborhood in the far east of the former West Berlin where most of the inhabitants are first- and second-generation Turkish migrants; he is returning to his family, after his father's death, to ask about his inheritance. Here we learn that Lola had been kicked out of the family and that Lola's lover, Billy, passes as straight when he is in Kreuzberg. The film portrays the immigrant neighborhood as socially controlling and repressive, which forces gay Turkish-Germans to find a liberated existence outside the ethnic neighborhood and family. Ultimately we see Lola floating dead in the River Spree, which divided the old East and West Berlin, and we learn that Lola was killed by her older brother, Osman, who raped her in the past in their apartment. In contrast to earlier homophobic films that posit the danger of violent homosexuality or anonymous homophobic violence in the public spaces of streets and parks, *Lola and Billy the Kid* suggests that murderous violence emerges from the domestic space of the patriarchal home.

This film juxtaposes traditional Turkish woman with the performative, queer, female identity of the main characters who are men cross-dressing as women. In one short but telling scene, Sherezade, one of the cross-dressing male characters, leaves her Kreuzberg apartment for good. Dressed provocatively in a miniskirt, she meets her neighbor, a traditionally represented Turkish woman in a long coat and headscarf, in the hallway. This woman has known Sherezade only as a man, and she is upset to see her as a woman and harasses her for her socially unacceptable behavior. Sherezade mocks her by telling her that in the past she cross-dressed as a man simply to ward off the hungry husbands of the neighborhood. She thanks the neighbor for being so kind, only to throw in her face on her way out: "But your husband was much nicer."

Lola and Billy the Kid is one of the films – like *Paris Is Burning*, discussed below in the case study – that integrates the city as the playground for queer desire with acknowledgement of discriminatory social, economic, and psychological structures. Importantly, the friends celebrate Lola's birthday on a merry-go-round in a playground. Performing the high-heeled femininity in the park in the middle of the night, the cityscape becomes a playground for performances of gender and sexual desire. Yet, in contrast to *Virgin Machine*, the light moments do not lead to a coherent utopia but instead point to utopian possibilities beyond the narrative of Lola's violent death, which reflects the violent social reality for gay and ethnic migrants in the city.

Case Study 8 **Jennie Livingston's *Paris Is Burning* (1988)**

Paris Is Burning portrays one of the most distinctive, creative, radical, highly organized, and original, urban subcultures in twentieth-century America, yet was a successful crossover to mainstream America, translating a subculture that was specifically located in the gay neighborhood of New York City into a text to be read and consumed across the country. The subculture portrayed in *Paris Is Burning* is organized in houses with playful and innovative names that are reminiscent of French couture houses, Labeija, Ninja, Pendavis, Saint Laurent, and Xtravaganza. These houses compete with each other and provide alternative families for their members. Every house has a father, mother, and children, but all of them are gay men of color, so the organization of the alternative families relies on performances that conflict with the traditional understanding of masculinity. Competition is acted out through modeling on an imaginary catwalk in complex categories of performance and dance-offs that include shading, reading, and vougeing. Again, the organizational structure foregrounds the competition in capitalism that is disavowed in a discourse around couture as a high art based on the inspiration of genius. As the subjects say in the film: "A house is a gay street gang; street fights take place at a ball by walking in a category."

Paris Is Burning portrays cross-dressing as cultural practice on the levels of gender, class, and profession. While these kinds of subcultures are possible only in the metropolitan area, they often take place in safe spaces within the city to safeguard the expression of alternative sexual identity. Thus, while the cityscape is shown in establishing shots to announce the urban terrain in which sexual desire expresses itself, and while often parks, streets, alleys, and public restrooms are sites for anonymous and illicit sexual encounters, films represent a presumably authentic subculture in spaces that are not entirely public.

In contrast to most of the other films discussed in this book *Paris Is Burning* is a documentary. Through its use of the genre the film situates itself in a complex history of anthropological, ethnographic, and educational films about ethnic, colonial, racial, class or sexual others. African-American public intellectual and feminist theorist bell hooks strongly criticized the film as inscribing a white, voyeuristic gaze by a lesbian filmmaker on a Black, poor practice. The film is structured according to generic documentary conventions organized by intertitles which include terms denoting subcultural practices and spaces that are then defined and illustrated for an audience presumed to be in need of illumination. The terms include: "Ball," "Legendary," "Realness," "House," "Shade," "Reading," "Mopping," and "Stunt," which are then explained by members of the houses in the form of talking heads.

The film celebrates the excesses of gay life in the face of the lack of material resources. The film's subjects repeatedly explain: "I never felt comfortable being poor. I always felt cheated." Others explain that the balls provide individuals with the fantasy of being a superstar; that some of them don't even eat, and sleep on the pier. The fantasy of being a superstar or a legend is validated by an alternative social structure that has a shared understanding of "legend" and "upcoming children." Here the legends are validated and created by their peers: sometimes legends are made through the representation of the most extravagant outfits, but at other times also through banal actions characteristic of everyday life in the "real" world: "going to school," "town and country," "military," "girls at a corner." The deciding category is realness: to be able to blend into the normal world organized according to traditional conceptions of masculinity and femininity.

The film employs cross-dressing to undermine our notion of the naturalness of gender through a very simple strategy. We see intercut into the narrative about the houses

continued

shots of "regular" people in New York City, crowds of people walking the street (similar to the shots of Joe and Rizzo in *Midnight Cowboy* or of *Shaft* walking down the street), a standard trope for the crowds walking around New York City, but suddenly the view of those unknown people changes. Whereas in *Midnight Cowboy* and *Shaft* the crowds serve to highlight the uniqueness or distinctiveness of the main character, the effect of the shots in *Paris Is Burning* is quite different: their normalcy suddenly seems staged and artificial: are the women really women and the men really men?

This is both a strategy by the subjects in the film and by the filmmaker and film itself. The informants explain the term "executive realness": "In real life you can't get the job because you don't have the education. Therefore you are showing the world: you can be an executive . . . your friends are telling you, you could be an executive." While a critical view of the film would suggest that the contradiction is obvious, namely that being an executive is not just the habitus and look and that the film reveals the shortsightedness and limitation of the film's subjects, a more positive view would credit the subjects with insight into American society and the ideology that covers over the material and economic inequality and discrimination that affects marginalized gay men of color. The repeated re-enactment of behaviors conventionally unmarked as middle-class and white sensitizes an audience to the performative dimension of naturalized identities (see J. Butler). When the presumably natural behaviors that produce whiteness, heterosexuality, femininity, masculinity, and middle–upper class are estranged, the film turns the tables on the underlying discriminatory structure of society and we are aligned with drag-queens of color, traditionally seen as marginal – outside the norm. Instead, normality, as it appears in the crowd scenes in the film and, supposedly surrounding viewers when they leave the movie-house, is estranged and appears as a constructed illusion of normality.

This kind of culture, however, had to disappear once it actually acquired fame because it existed in the curious dialectic of hypervisibility and invisibility that marks so many subcultures. Several of the characters state at the end of the film that the balls came to be toned down, becoming long, drawn out, and boring; that they miss the street element; that "the scene" changed drastically, and that New York is not the same anymore. The subculture that was dependent on the city but was simultaneously underground disappeared with the cleaning up of New York City and the mainstreaming, cooptation, and commodification of gay culture. Thus, the original subversive quality, and the pleasure associated with it, has disappeared.

Paris Is Burning portrays the erotics of sartorial pleasure. In contrast, sex acts, similar to those of *Lola and Billy the Kid* are the source of economic sustainment: prostitution as a means to an end that is found in the city but is bypassed by the subjects in the film. They hint at it but they do not want to discuss it. The film endorses the fantasy that its subjects advance. Only at the end do we learn that Venus Xtravaganza, one of the characters featured most in the film, who is charming, kind, childlike and particularly open about the desire to be a rich, white girl so that she would be spoilt, was in reality strangled and left for dead under a bed prior to the film's release. The violence against those who do not conform is not represented and investigated by the film, which instead celebrates the subversive creativity of the marginalized subculture.

In summary, *Paris Is Burning*, is a city film, in which the locale enables a practice that nevertheless cannot be undertaken entirely in the open, and so remains an underground cultural phenomenon. The film exposes an unacknowledged side of New York, a city that stands for the urban melting-pot, a notion that is rigorously questioned by the representation of those who are excluded. At the same time, New York City represents a destination for those who run away from home trying to find an alternative to the traditional family when it is experienced as oppressive. Only in such a city can such elaborate alternative social structures develop, yet when it becomes part of the official selling point of the city (such as in San Francisco or Berlin), it loses its subversive quality.

Paris Is Burning celebrates a subculture and simultaneously documents and archives it at the moment of its disappearance. The cultural practice at its center would otherwise not be marked in the urban space, if it wasn't for this film. No plaque, no museum, no statue would celebrate the memory of vougeing. As a lived urban practice, these gay men of color share this fate with other marginalized urban groups, such as squatters. While the film may be justifiably criticized for allowing a voyeuristic, ethnographic look at an otherwise hidden subcultural practice, it must also be accorded its rightful place as a document of an otherwise invisible urban practice that created its own urban space that has since been destroyed and otherwise forgotten. And in that sense, the documentary film, in which the subjects talk, re-enact, and act out may be the most appropriate medium in its rehearsal of their cultural practices at that moment and in that place. *Looking for Langston* relies on, if nothing else, the published, written, painted, and sculptured artistic artefacts which capture the moment of the intersection of urban place of Harlem, its cultural production of the Harlem Renaissance, and the gay sexual desire associated with it. The collective cultural practice of vougeing, going

continued

to a ball, has all but disappeared with the sanitizing of New York City. *Paris Is Burning* captures a moment of radical change: from sexual activity to sartorial pleasure (related to the AIDS epidemic), from marginalization in metropolitan areas to sanitizing urban planning. While documenting these parallel urban and socio-sexual changes, Jennifer Livingston has also given her subjects something akin to their desire: a legend of urban resistance with gold, glimmer, and high heels.

Further reading

David Bell, Jon Binnie, Ruth Holliday, Robyn Longhurst, and Robin Peace (2001) *Pleasure Zones: Bodies, Cities, Spaces*, New York: Syracuse University Press. The collection brings together different kinds of case studies about the relation of sexuality to urban spaces from varied methodological approaches that can be applied to the cinematic representation of space in contemporary queer film.

Martha Gever, John Greyson, and Pratibha Parmar (eds) (1993) *Queer Looks: Perspectives on Lesbian and Gay Film and Video*, New York: Routledge. A volume that collects the diverse views of scholars on the explosion of queer cinema in the early 1990s.

Gordon Brent Ingram, Anne-Marie Bouthillette, and Yolanda Retter (eds) (1997) *Queers in Space: Communities, Public Places, Sites of Resistance*, Seattle, WA: Bay Press. This is an extensive collection of contributions from scholars in geography and queer studies that tie queer experience to places, sites, and architectures.

Richard Dyer (1990) *Now You See It: Studies on Lesbian and Gay Film*, New York: Routledge. This historical overview is an account of gay and lesbian cinema from Weimar Republic onward.

Essential viewing

Linda Goode Bryant and Laura Poitras. *Flag Wars* (2003)

Isaac Julien. *Looking for Langston* (1988)

Jennie Livingston. *Paris Is Burning* (1988)

Takehiro Nakajima. *Okoge* (1992)

Richard Oswald. *Different from the Others* (1919)

9 The global city and cities in globalization

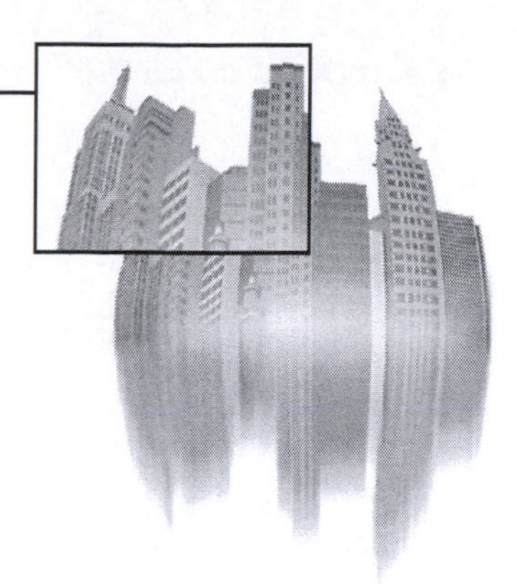

Globalization is a process that generates contradictory spaces, characterized by contestation, internal differentiation, and continuous border crossings. The global city is emblematic of this condition.

Saskia Sassen

Learning objectives

- Understand the dynamics of globalization
- Comprehend the changes of cities in the context of globalization, particularly in regard to global networks and the changing meaning of topography
- Examine the characteristics of funding, production, distributions, themes, and narratives in transnational cinema
- Be aware of thematic clusters of transnational cinema, such as migration, illegal labor, sex traffic, transnational love stories

Introduction

This chapter addresses the relationship between cinema, the city, and globalization. In order to tease out the links that connect film to the representation of urban centers in the era of globalization this chapter outlines the current understanding of globalization, highlighting the role of cultural production in global circulation, on the one hand, and the status of the city, on the other. The chapter then accounts for different cinematic responses to globalization and discusses selected films addressing the ways in which globalization engenders particular narratives and constructs a global city. The global city has become "a central location for capital exchange – a central node in a vast, multinational network of capital and labor flow" (Oren 53). Transnational cinema addresses the global flow of labor and culture in its representations of metropolitan areas that have emerged in globalization. Global cities provide settings for narratives about migration, but the cinematic representation of global cities also offers new global versions of older tropes associated with the city, such as alienation, now reflected in the representation of tourists, business travelers, and the displacement of migrants within global networks. The chapter also includes a discussion of a video that in its aesthetic structure reflects this kind of alienation in global sex-trafficking and ties it to accounts of global geography.

What is globalization?

Globalization "denotes the processes through which sovereign national actors are criss-crossed and undermined by transnational actors with varying prospects for power, orientations, identities, and networks" (Beck 2000b: 11). Increasingly the movement around the globe of capital and products is enabled so that cultural products are available in countries other than their nation of origin (Short and Kim 3). John Rennie Short and Yeong-Hyun Kim differentiate between economic globalization, cultural globalization, and political globalization, but maintain that globalization's greatest effect is in the sphere of finance (3–4). The power of the nation–state has been weakened, and market-driven agents compete with national governments in the processes of transnational trade. Short and Kim correctly observe that the analysis of globalization has focused on economics and that issues of global culture have been subordinated to discussions about economic and social processes of globalization. Global cinema relies on multinational funding, production, and distribution, in contrast to national cinema, where national funding, national culture, and a national audience were in place and taken for granted. Paradoxically, it is the development of globalization that has led to a more thorough interest in the conditions of national cinema.

Transnational cinema

Film is part of the global flow of cultural products *and* also represents globalization visually and narratively. A detailed outline of transnational funding and distribution structures would exceed the parameters of this chapter, but paradoxically, studies of that kind have appeared primarily in national film studies contexts (see, for example, Halle). Transnational processes of production, such as co-production as part of bilateral agreements or cultural agreements like the European Union and the international reception through wide circulation of films, offer opportunities for creative exchange beyond national frameworks of production, reception, and interpretation in a global world. While some see the effect of globalization on cinema as an increase of the hegemony of Hollywood at the expense of all other national cinemas, I propose a more positive view of globalization, emphasizing the creative possibilities of cinematic exchange.

The relationship between place and culture has changed; culture has in many respects become deterritorialized, which means that cultural artefacts do not have singular places of origination to which they are beholden. Globalization is an uneven phenomenon, and the distinct national film industries have different kinds of access to production funds and world markets. Hollywood films dominate, while smaller film industries have a harder time being produced and distributed, yet all film industries have undergone change under globalization. Now filmmakers from national cinemas are also influenced by global visual culture. Chapter 7 has illustrated this shift with films that depict African-American ghettos in the context of the USA to the circulation of the transnational ghetto film that is exported from the USA to other countries, localized, and then reproduced and globally recirculated as national product from South Africa (*Tsotsi*), Brazil (*City of God*), or Germany (Fatih Akın's *Short Sharp Shock*, 1998). Filmgoers consume films from all over the world, but particularly Hollywood cinema. Filmmakers' narratives, and their thematic and aesthetic sensibilities are influenced by films from outside their national cinemas. More often filmmakers are trained abroad and multi-national funding is tied to multi-national narratives, forcing filmmakers to tell stories about border-crossings.

Films about globalization address a range of issues. Documentary films address the effects of globalization on the Jamaican economy, as in Stephanie Black's *Life and Debt* (2001) and Raoul Peck's *Profit and Nothing But! Or Impolite Thoughts on the Class Struggle* (2001); the dreams and hopes of crossing the border from Mexico to the US, as in Chantal Akerman's *From the Other Side* (2002); and the killing of women in Ciudad Juarez, a Mexican–American border town with *maquiladoras*, foreign-owned assembly plants, as in Lordes Portillo's *Senorita Extraviada* (2001), and Ursula Biemann's *Performing the Border* (1999). Feature films include narrativizations of multiple border crossings, including the Polish–German border, as in Hans-Christian Schmid's *Lights* (2003); stories about war refugees in makeshift refugee camps, as in Bahman Ghobadi's *Turtles Can Fly* (2004); traveling businessmen who encounter foreign landscapes, as in Sue Brooks's *Japanese Story* (2003); and displaced tourists, as in Sofia Coppola's *Lost in Translation* (2003). These films highlight different kinds of cinematic spaces than we have encountered in the preceding examples: the borderlands, *maquiladoras*, the makeshift refugee camp, and the global city. It is important to differentiate between films that are produced for a global market and independently produced films that both address globalization and circulate in global networks, even if in less powerful ones.

Urban centers in a globalized world

Globalization manifests itself in cities, especially in the large global metropol. It affects "major metropolitan centers of international finance and business – cities like Tokyo, New York, Paris, Hong Kong, and Berlin" on the one hand, but also the "spaces and places between and beneath major urban centers" on the other (Petro and Krause 2). Short and Kim argue that the unevenness of globalization leads to "[m]any developing countries in Africa and Latin America [with] extremes of wealth and poverty . . . coexisting in the same city" (8). An extraordinary illustration of this is to be found in Ra'anan Alexandrowicz's film *James' Journey to Jerusalem* (2003), which is this chapter's case study. The migrants and illegal laborers portrayed in the film travel through and into the wealth of the city, while they live in a neighborhood that lacks any kind of infrastructure.

Petro and Krause focus on the ways in which "global networks" are "experienced locally," which can mean in small or large cities as well as rural spaces (2). In this digital age, however, Saskia Sassen argues that the "built topography" is "increasingly inadequate" to represent cities (2003: 15). The digitalization of culture exceeds the traditional accounts of topographic representation that cannot capture the new forms that power takes in globalization. The topographic representation of poor areas of a city portrays only the physical aspects of poverty, and Sassen suggests that the "dominant interpretation" of globalization and digitalization posits "an absolute disembedding from the material world," in short "that place no longer matters" (2003: 16). Against these kinds of simplified understandings of globalization, Sassen argues for "new types of spatializations of power" (2003: 17). Most visual texts discussed in this chapter are traditional films and as such are indebted to the realistic representation of space and place. They show that the distribution of power manifests itself in the spaces of globalization.

Remote Sensing

I begin here with a more radical example of a visual response to globalization that is not indebted to the realistic recreation of a cinematic city within a narrative of global movement, but rather appropriates digital culture for an avant-garde video that places cities as part of migratory routes of women in the transnational sex trade. Ursula Biemann's video essay *Remote Sensing* (2001) contrasts the digital depiction of routing and rerouting bodies around the world with the practice of geographic remote sensing that allows for both global surveillance, on the one hand, and the fantasy of a borderless, global world, on the other. *Remote Sensing* brings together several topics that tend to be segregated in discourses on globalization: sexuality and economics, national histories and transnational memories, the national and the global, scientific accounts of geography and the experience of global movement and place. It connects different accounts of the routes that women travel in the global sex industry using NASA satellite images.

The migratory routes are recounted by individual women, as well as contextualized by Czech, Indian, Thai, Swiss, and Filipina activists, primarily from local and global organizations against the traffick in women, and the filmmaker in a voice-over. Repeatedly routes are shown in the form of computer-screen texts akin to air travel itineraries layered over images. The video relies on multiple sources of information, which include facts and numbers, narrations of movement, and meditations on places, and the NASA satellite images, which are emblematic of a global geography and, since they are accepted as highly scientific, create an imaginary world without visual borders. The NASA images thus reflect the space–time compression of globalization as described by Andreas Huyssen: "the experiential dimension of space has shrunk as a result of modern means of transportation and communication" (2003: 1, see also Harvey 240).

Remote Sensing captures the complexity of women's multiple and multi-directional migratory routes without claiming to be a comprehensive account the global trade of sex, false adoption, and marriage. The movement described is not limited to movement from an imagined periphery to an imagined center, such as from one country to Switzerland, Germany, or Europe for that matter, which would reproduce a Eurocentric worldview. Instead, the video tracks women traveling from Thailand to Tokyo, Russia to Korea and Israel, from Cambodia, Laos, and Burma/Myanmar to Thailand, from Eastern to Western Europe, from Nigeria to Germany, Vietnam to China, and from Latin America to the USA. To emphasize the multi-national multi-directionality, a section of the video entitled "Filipinas in Nigeria: A Case of Re-Routing," highlights the story of two women who are involuntarily trafficked to Nigeria and take years to return home to the Philippines.

The two women, Arlene Banson and Nilda Vibar, tell their story: lured by the promises of Bernadette and Wolfgang Stromberg, presumably two German nationals, to work in a German restaurant, they find themselves in a brothel in Nigeria, forced to service Chinese customers. The screen – split sometimes in three, sometimes in four sections – shows us the two Filipinas telling their stories in one of the top sections while a satellite image of the Philippines and/or Nigeria moves across the other sections of the screen. The split screen reflects the complexity of globalization, in which the categories of nationality, gender, and ethnicity function but are forcibly separated from their, presumed essential, relation to each other. The multi-national and multi-directional movements are captured when the trafficking routes are layered over the NASA satellite images, for example: Metro Manila → Oriental Club Lagos/Nigeria → Nightclub Lome/Togo → Nostalgia Club Larnaca/Cyprus → Metro Manila. Whereas in documentaries the commentary interprets the simultaneous image, in Biemann's video the sound and image diverge to reflect an asynchronous experience of time and place characteristic of globalization in general, but more importantly characteristic of the experience of the particular women portrayed in this video.

The mere depiction of visible reality has become insufficient, especially when addressing the clandestine realities of globalization, such as the traffick in people, illegal migration, and global sex work. The self-reflexive voice-over asks: "How to shoot a clandestine life?" That self-reflexivity points towards the filmmaker, Biemann, and to us, the audience, urging us to question the visible when we are shown seemingly innocuous shots – of men, women, doors with numbers, images whose hidden meaning can be understood only through context provided by the voice-over. The commentary reflects on the conditions of secrecy without sacrificing the explicit naming of sexual negotiations and transactions that motivate the clandestine existence of women in the sex trade in the first place: "Locked up in tiny rooms . . . guarded step by step, number by number, trick by trick . . ." While seemingly innocuous landscapes are exposed as sites of sex trade, we are also shown images of self-confident women enjoying themselves. For example, *Remote Sensing* cross-cuts several shots of women riding together on scooters through an urban landscape, smiling at the camera. In Ulrich Beck's account of globalization as a "second age of modernity," gender does not play a role (2000b: 79), but *Remote Sensing* shows us a gendered modernity brought on by globalization when women experience upward mobility and changing gender roles by moving to richer urban and metropolitan sites with possibilities of increasing their income.

Agency is also accorded to women when they relate memories of their journeys. Andreas Huyssen suggests that cultural globalization engenders a significant shift from "national history within borders" to "memory without borders" (2003: 4). But how do we conceptualize the paradox that the global cultural production of "memories without borders" is increasingly shaped by the trauma and/or the impossibility of border crossings? Biemann's video complicates the juxtaposition of "national history within borders" and "memory without borders" by illustrating how national and international histories, particularly those of military conflict, create national sex industries which, in the context of global inequalities, produce transnational trafficking of women's bodies. Thus, *Remote Sensing* integrates two aspects of globalization that seem to be at odds with one another: the time–space compression, which makes borders and nations seem to disappear, and the traffick in bodies, which move through national spaces and thus highlight the space of the nation and its borders. Biemann's video creates memories without borders that circulate globally and are not bound by a single national imaginary. This video essay engages the digitalizing effects of globalization with its local material manifestations.

Global cinematic topographies of the city

Transnational films integrate the topography of metropolitan areas with the transnational movement of characters. The flow of finances hardly provides material that can be translated into the visual medium of film, so transnational cinema shows metropolitan areas in different countries connected by the various aspects of globalization: labor migration, international tourism, transnational commodification, postcolonialism, transnational education, transnational capital, and the transnational sale of body parts. Connected by narrative topics and representations of the city as a space of alienation and solidarity, the films show the visible effects of globalization and its subcultural and submerged illegal underside.

Films attempting to capture the manifestations of global flows in the visual representation of concrete space and the built environment negotiate the relationship between the local and the global. Sassen explains that what we might experience as local – for example in the built environment that surrounds us – functions in reality as a "microenvironment with global span," because it is globally connected through different kinds of networks (2003: 20). She goes on to call these environments a "localized entity," one which is experienced it its local place and captured in topography but it is connected globally. Films such as Stephen Frears's *Dirty Pretty Things* (2002) and *James' Journey to Jerusalem* (2003) represent the built environment of material sites criss-crossed by movement. *Dirty Pretty Things* employs the setting of a hotel as the place where migrants' bodies are harvested and their organs sold. The film transforms the local space into a disturbingly disconnected, decontextualized hotel in an unnamed city to capture the simultaneous local manifestation of globalization and the increasing independence of locations form their local context. This alienation is different from that of Weimar cinema and film noir, in which the cinematic topography creates a whole city. This new global cinema answers the question of how to cinematically capture the relationship of local and global, as described by Sassen: "The local now transacts directly with the global – the global installs itself in locals, and the global is itself constituted through a multiplicity of locals" (2003: 24–5).

Ethnoscapes of tourism

The globalization theorist Arjun Appadurai argues that it is "electronic mediation and mass migration" that particularly mark the present moment of globalization and enable the "work of imagination" that we see reflected in global cinema (4). To discuss the shifting spaces, real and imagined, that constitute the global world, he coined the term "ethnoscape," defined as "the landscape of persons who constitute the shifting world in which we live," including "tourists, immigrants, refugees, exiles, guest workers" (33). These kinds of ethnoscapes are often recreated in the cinematic representation of global cities inhabited by people moving transnationally into, out of, and through them.

Sofia Coppola's *Lost in Translation* (2003) creates such an ethnoscape of upscale tourist and business travelers in Tokyo, Japan. The film's main character, American actor Bob Harris, visits Tokyo to shoot a commercial. In his hotel he repeatedly runs into fellow American Charlotte who is staying there with her husband, a photographer. Charlotte, who is significantly younger than Bob, is alone in the hotel for long stretches of time while her husband is out on photography shoots. The narrative is loosely tied together by Bob and Charlotte meeting and spending time together in the city. Most of the film shows episodes of activities that tourists engage in when they are staying in hotels, such as sitting in the bar, listening to jazz, swimming in the hotel pool, and listening to tapes in the hotel room. On occasions, they venture outside, sometimes alone, sometimes together, sometimes at a voyeuristic distance to the surroundings, sometimes engaging with friends. At the end of the film, they both depart.

The primary setting of the film is an upscale hotel in Tokyo, which offers a hybrid space of local Japanese customs, such as the formal introductions at which business cards are exchanged, and the transnational culture, influenced by American and European traditions that we see in the hotel bar. The film repeatedly returns to the bar, where entertainment is provided by a pianist and a singer who perform smooth jazz, the international soundtrack familiar from airports, elevators, malls, and upscale hotels and bars around the world, expressing the homogenized global culture that circulates particularly in the high-end world of global business travel. The bar is populated by Japanese, Americans, and other nationally unmarked business people and tourists, yet their behaviors do not attest to any cultural differences. They all sit quietly at their tables or at the bar and drink liquor and smoke cigars.

The nationally undefined space of the hotel bar is surrounded by the night cityscape of Tokyo. Dark and quiet, customers sit in low chairs by windows that offer a panoramic view of Tokyo at night. The city is always kept at a distance, visible through hotel room windows, as the backdrop of the hotel bar, or moving past Bob Harris when he is driven in a cab through the city from the airport to the hotel. The film does not capture Tokyo or the adventures of a traveler who immerses himself or herself in a foreign culture – in film often associated with self-discovery – but rather the alienation of the tourist. The film's title, *Lost in Translation*, refers to the confusion that ensues when one loses one's foothold in the security of one's own language, and the film also captures the bodily confusion and problems of adapting to long-distance travel: Bob and Charlotte meet because they cannot sleep, and their encounter with Tokyo is a sleep-deprived, sleep-walking activity.

When the characters venture into the city, the narrative retains its episodic quality. They make no lasting connections, instead finding themselves at bars, doing karaoke, and walking through the crowds. The film presents no conversation that is not superficial and no connection that goes beyond curious inquiries expressed in limited English. The only connection is one that takes place between Bob and Charlotte, the two Americans who are displaced and alienated from their loved ones and their home. Bob receives faxes and express packages from his wife, who is renovating their house, requesting that he make decisions about shelves and carpets. The speed of communication and the presence of questions about home reflect the space–time compression enabled by electronic communication and characteristic of globalization. But it seems that the space–time compressions in the attempted communications by Bob Harris's absent wife only heighten the disjointed nature of Bob's experience of there and here, home and away, instead of overcoming the distance that is, the film implies, not only geographical.

The film captures the transnational space of the traveling business class. In so doing, however, it reproduces the septic gaze on foreign nations and cities in which airports, airport shuttles, high-end hotels, and restaurants function. It never allows any deeper understanding of Japanese or Tokyo culture and in that process relies not only on the dominance of Hollywood to produce images and representations of other countries but also on older stereotypes of the city in general as a space of random, superficial encounters and Asianness as indecipherable for Westerners. The film's limited dialogue and action create a sense of paralysis and sleep-walking for the audience, with no anchoring space except for the hotel bar. Ultimately the Japanese seem shallow and curious, perceived only through the alienated gaze of two displaced characters, a lonely young woman and an equally lost older man who happen to share a common language and culture. Taking into account that English is the dominant language of globalization, the film's statement about the possible productive outcomes of global exchange is rather hopeless, positing meaningful human contact only in English.

Migration: legal and illegal

Legal and illegal migration is a pre-eminent topic addressed in transnational cinema. In France films identified by Peter Bloom as *beur cinèma* show stories about migrants on the continuum of illegal and legal migration and residence in Marseilles and Paris. Such films – consider Madhi Charel's *Tea in the Harem* (1985), Karim Dridi's *Bye-Bye* (1995), and Mathieu Kassovitz's *Café au lait* (1993) and *Hate* (1995) – portray life among North African and Caribbean migrants in French cities, sometimes moving back and forth between the country of origin and the host country. Several of these films overlap with the chapter on the ghetto and barrio, evidence that globalization is not an entirely new phenomenon but rather a radical break that has led to an intensification of phenomena, such as migration. Hamid Naficy has proposed the term "accented cinema" for those kinds of films that share aesthetic features in conjunction with narratives of migration, exile, and diaspora. Accented cinema in Britain includes Damien O'Dennell's *East Is East* (1999) and Gurinder Chadha's *Bend It Like Beckham* (2002), which addresses the Indian and Pakistani communities in Britain. Germany has also seen an explosion of Turkish–German cinema in the last decade, including Fatih Akın's *Short Sharp Shock* (1998), *In July* (2000), *Solino* (2002), and *Head-On* (2004), and Thomas Arslan's *Brothers and Sisters* (1997). Sassen, in contrast to many other economic theorists of globalization, emphasizes the alternative use that minorities can make of the city: "Those who lack power – those who are disadvantaged, who are outsiders, who are members of minorities that have been subjected to discrimination – can gain *presence* in global cities, presence *vis-à-vis* power and presence *vis-à-vis* each other" (2003: 25).

Case Study 9 **Ra'anan Alexandrowicz's *James' Journey to Jerusalem* (2004)**

Ra'anan Alexandrowicz's *James' Journey to Jerusalem* (2004) is not the best-known of the films that address globalization, but it is widely available and distributed around the world. It captures the fine line that divides legal from illegal migration and situates illegal labor in the unlikely context of a religious pilgrimage, a surprising point of departure, particularly in relationship to Israeli cinema. The film's charm is associated primarily with the main character, James, but also results from its light-hearted portrayal of a serious issue that allows the film to be billed a "new comedy from Israel." It thus belongs to the genre of immigration comedy

– like Dani Levy's *I Was on Mars* (1992), Jan Hrebejk's *Up and Down* (2004), *East Is East*, and *Bend It Like Beckham*. *James' Journey to Jerusalem* moves beyond the dominance of Hollywood and portrays the pervasiveness of globalization worldwide, describing a journey from South Africa to Israel and partaking in the account of the global network of migration in the periphery of the global capital of filmmaking.

James is a young Christian from Africa who is sent by his village on a pilgrimage to Jerusalem before he can become a priest. The film is framed in the opening and at the end by a song and paintings that tell the story of James's pilgrimage to Jerusalem and depicts globalization in a premodern form of storytelling. At the border he is arrested as an illegal labor migrant. Shimi, who employs and houses illegal laborers in Tel Aviv, buys James out of jail, takes his passport, and puts him in an apartment with a group of illegal immigrants. Shimi makes James work for his own father, Mr Salah, who is a widower and Holocaust survivor. Mr Salah insists on living in his small house with his garden and the memory of his deceased wife, while his son Shimi wants him to sell his place to a developer so that he can make money from the land value. While James works for him, Mr Salah teaches him a few Hebrew words and concepts about advancing economically through the exploitation of others instead of letting others take advantage of him. James also works for Shimi's wife and her female friends, who enjoy not only the cheap labor he provides but also looking at the attractive, young, African man. On Sunday James attends church, and the priest also asks him for money for the congregation. In order to buy back his passport – for the money that he owes Shimi for getting him out of jail and to donate money to his church – James begins to work long hours and in addition to use his fellow laborers to work for him, making money from their labor behind Shimi's back. In the evenings he escapes to the city with his friends and enjoys looking at and buying the goods that are available in malls. At a party thrown by Shimi to celebrate the sale, finally, of his father's land, James and Shimi get into an altercation, and James is arrested, and taken to a Jerusalem jail because of his illegal status. When James realizes on the way that the prison bus is driving past a vantage point on the city popular with tourists, he asks to have a photo taken of him in the foreground of the city of his destination. The film then cuts to the framing device of an African song and a sequence of paintings, both of which portray James's journey to Jerusalem.

Aesthetically the film reflects the juxtaposition of the innocent James, with his mythic origin, and the bustling, industrial, corrupt city by portraying his original journey from the unnamed African village to Israel via naïve paintings, accompanied by

continued

a song that also includes the film's credits. The paintings capture folk art by being painted on different pieces of wood; the first image nevertheless appears as a triptych, painted on three pieces of wood. The final painting is a beautiful portrayal of the golden city, which then fades into a cinematic image. When the camera pulls away from the beautiful golden city in an over-the-shoulder shot, it is revealed to be a poster of Jerusalem that James is perusing in admiration. Dressed in traditional African clothes, he is standing in a bureaucratic room that is bare but adorned with posters of Israel.

The movement from traditional paintings to film to poster and back to film is repeated at the end of the film when James is transported to the jail in Jerusalem because the prison in Tel Aviv is filled with illegal immigrants. He is transported by two police officers and suddenly understands, when he sees the signs on the highway, that they are on their way to Jerusalem. He strains to look out of the window through the metal grid and begs to be photographed with Jerusalem in the background for the people in his village. The officers comply, and the final shot of the narrative is James smiling into the camera, doubled by the photo camera and the film camera, with a smiling police officer next to him, and the film changes to the paintings from the opening sequences, framing the film in an oral story through the words of the song: "A long journey to Zion . . . to the promised land beyond the seas." The song provides the background information about "a faraway village," where they pick their best young man and send him to Zion so that he can return and tell about it. The song ends with a jubilant refrain: "Jerusalem, you are our destiny."

James' Journey to Jerusalem takes much of its energy from the imaginary contrast and conflation of the idea of a religious and pre-industrial pilgrimage and global migration for labor. Its irony lies in the fact that both narratives lead to Jerusalem, and James ends up with a photo to take home. The film employs his figure to criticize the discrimination and exploitation of national and regional economies, individual national and ethnic groups, as well as individuals. Its critique extends to the visual registers of recognition and the limited visual access the immigrants have to the city. The border police do not believe James's story because they perceive Africans to be illegal immigrants. The film also criticizes the forms of exploitation where formal and informal sectors intersect: for example, Shimi works with the police, who participate in the illegal exploitation of workers. *James' Journey to Jerusalem* also portrays the restricted access the migrants have to the city in emphasizing their partial view out of the van in which they are driven through the city. Most shots of the city rely on a restricted view out of the tarpaulin on the little truck with which the men are carted through the city to and from work.

An exception to this limited view of the city is provided by the excursion that James undertakes with his friend and fellow migrant Skomboze to downtown Tel Aviv, which is contrasted to the underground existence of the migrant workers in the substandard apartments, the father's shack on the outskirts of the city, and the new buildings that are being constructed in an empty wasteland. After their first paycheck, James and Skomboze take the downtown bus. The film shows the movement between the different parts of the city, from the migrants' apartment to the distant, affluent downtown that is the space of consumption. Without dialogue we see James and Skomboze in the mall, looking at clothes, electronics, and everyday goods and shoppers from their point of view. They are dressed like everyone else in jeans and bright t-shirts and enjoy an unrestricted view of the goods of consumption. The film shows us the bright, artificial colors in big ads in contrast to the small garden in which James works for Mr Salah in real daylight. Their unrestricted view and the unrestricted view of them, in contrast to their daily existence in the hidden illegal sector of the city, are emphasized when they take the glass elevator in the mall. Repeated shots show the exchange of money and the goods themselves, cell phones in all colors, rows of televisions. Since, as an audience, we visually share James's deprivation during the working day, we also share the sudden visual excess of bright and modern consumer goods. The last shot of their first excursion shows them looking down into the mall and James quoting the bible: "As it is written . . . a land flowing with milk and honey," marking his confusion about his pilgrimage with the excess of goods.

The trip into the city is accompanied by an establishing shot of downtown Tel Aviv in contrast to the lack of sense of location of the apartment that houses the illegal immigrants. The film emphasizes the sections of the city that look and function entirely differently. Sassen describes these different areas of the global city and locates what she labels "the informal sector," defined as "those income-generating activities occurring outside the state's regulatory framework that have analogs within that framework" (2003: 153). She suggests that research has paid attention to the "informal sector" in part to "stop excess migration to the cities," and that earlier explanations of the growth of the informal sector in global cities in the industrial West have been explained "as a result of immigration from the Third World and the replication here of survival strategies typical of the home countries of migrant worker" (1998: 154). Sassen argues against this theoretical position, and instead suggests that "the opportunities may well be a structured outcome of the composition of advanced economies" (154). The film supports her analysis. James reveals to Mr Salah that in Africa he was a farmer, and during the film he turns his little garden from a wasteland into a green oasis. It was not James's

continued

idea to work as a gardener; rather it was Shimi's organization that positioned him in his dad's yard.

The film uses the magical character of James as a foil for criticism of Israeli society. It is precisely James's childlike naïveté that allows the film a presumably innocent look at Israeli capitalism and highlights its flaws through the seemingly neutral eyes of an outsider. Also, the film portrays capitalism as a particularly Israeli phenomenon by having Mr Salah initiate James through language lessons. *James' Journey to Jerusalem* shows that his economic success is achieved by his exploitation of his fellow migrants, which changes his appearance and outlook on life. At the end, however, he cannot resolve his economic success because the Israelis still see him as a servant, and he throws his hard-earned money in their faces and decides to continue his journey to Jerusalem. Ironically, just then Shimi has him arrested, and he is taken to Jerusalem as a prisoner.

Tasha G. Oren explains:

> Popular narratives repeatedly enact the journey from the small town or village into the bustling city as a mythic journey into adulthood, as entry into a grand, aloof, and often ruthless world that simultaneously promises riches and threatens annihilation: material success and greed, diversity and alienation, self-knowledge and spiritual corruption, eccentricity, freedom, and Big-Time Evil.
>
> (52)

James' Journey to Jerusalem pits "Western commercial culture against an organic local that is pure, natural, and exotic" (55). The film presupposes a pre-industrial African identity untouched by globalization, which is an idealist construction and does not reflect reality. Most economies of Third World countries are shaped by globalization just as much as First World economies, and their citizens are often more aware of the effects and the disjunctures produced by processes of globalization. Sadly, the film thus reproduces the stereotype of the untouched and unenlightened migrant subject. But by also associating James with saintly qualities and otherworldly skills and magic (he can always roll the dice and get a six in backgammon), the film's narrative also exceeds the realistic and economic accounts of globalization.

Further reading

Ulrich Beck (2000b) *What Is Globalization?* Oxford: Blackwell, 2000. Very good introduction to globalization, written from a German sociological perspective, which means that one of the important concerns is the future of the social welfare state.

Elizabeth Ezra and Terry Rowden (eds) (2006) *Transnational Cinema: The Film Reader*, London: Routledge. A collection of important essays about the nature of transnational and minority cinema. The essays advance different and conflicting positions about the nature of global cinema. (see also "Further Reading" of Introduction).

Hamid Naficy (2001) *An Accented Cinema: Exilic and Diasporic Filmmaking*, Princeton, NJ: Princeton University Press. Hamid Naficy argues that exilic–diasporic cinema creates a set of aesthetic and thematic characteristics that are not necessarily a product of globalization. Instead, his approach offers an alternative model to discussions about global cinema.

Patrice Petro and Linda Krause (eds) (2003) *Global Cities: Cinema, Architecture, and Urbanism in a Digital Age*, New Brunswick, NJ: Rutgers University Press. A collection of essays addressing different topics, including memory, architecture, and cinema in cities, some of which have been addressed in this book, some of which go beyond the themes discussed here.

John Rennie Short and Yong-Hyun Kim (1999) *Globalization and the City*, London and New York: Longman. This co-authored volume gives a good overview over the different aspects of globalization in relation to cities around the world.

Essential viewing

Ra'anan Alexandrowicz. *James' Journey to Jerusalem* (2004)

Shu Lea Cheang. *Fresh Kill* (1994)

Stephen Frears. *Dirty Pretty Things* (2002)

Cédric Klapisch. *Euro Pudding* (2002)

Hans-Christian Schmid. *Lights* (2003)

Conclusion: from the "train effect" to the "*favela* effect" – how to do further research

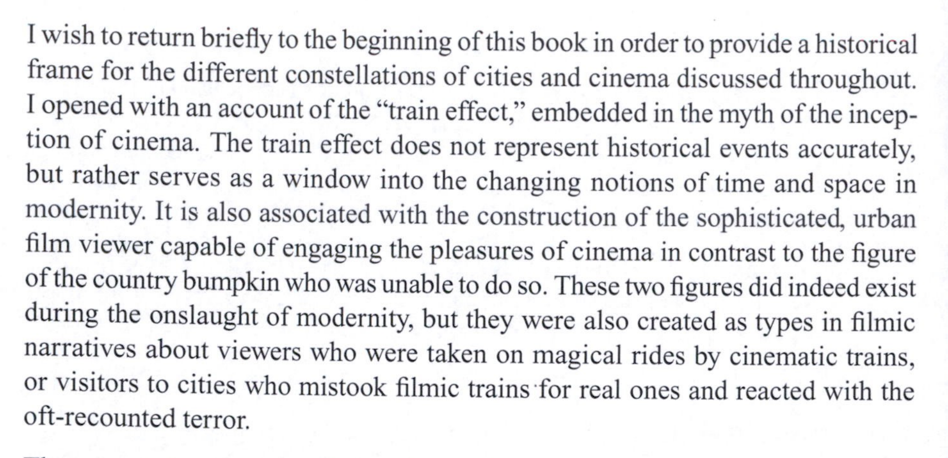

I wish to return briefly to the beginning of this book in order to provide a historical frame for the different constellations of cities and cinema discussed throughout. I opened with an account of the "train effect," embedded in the myth of the inception of cinema. The train effect does not represent historical events accurately, but rather serves as a window into the changing notions of time and space in modernity. It is also associated with the construction of the sophisticated, urban film viewer capable of engaging the pleasures of cinema in contrast to the figure of the country bumpkin who was unable to do so. These two figures did indeed exist during the onslaught of modernity, but they were also created as types in filmic narratives about viewers who were taken on magical rides by cinematic trains, or visitors to cities who mistook filmic trains for real ones and reacted with the oft-recounted terror.

These two types not only reflected the need to create new ways of seeing but were also put to the service of selling the new invention of moving pictures by celebrating the sophisticated urbanite as the ideal film consumer. In sum, the train effect encapsulates the changes in perceptions of time and space in modernity, the cognitive adjustments to the medial representation of reality, and the changing patterns of traffic associated with automobiles and trains, all of which were part of the economic matrix of the emerging film industry. The train effect functions as shorthand for the particular configuration of social reality, subjectivity, cognitive perception, economics, and medial representation that took place in the context of nation and modernity.

Even after more than 100 years, the relationship between cinema and cinema-goers remains a dynamic process. Now for example, at the turn of the twenty-first century, transnational cinematic exchange and postmodern culture have produced what I call the "*favela* effect." *Favelas* are the Brazilian equivalent of shantytowns, and are to be found at the periphery of most Brazilian cities. In contrast to other slum areas in Third World countries that result from rural-to-urban migration, *favelas* in Brazil came about when groups of displaced people settled outside of the major cities. They are traditionally built on hillsides by the inhabitants themselves, who often create informal infrastructures such as radio stations and electricity supplies.

Several important Brazilian films, such as Fernando Merielles's Oscar-nominated *City of God* (2002), which was set in a *favela*, have disseminated representations of these shantytowns to a global audience. In 2007, the journalist Andrew Downie reported on tourists from Germany, Britain, and the USA visiting and settling in Rio de Janeiro's *favelas* in order to experience "the real Brazil," insisting on living the authentic life in contrast to middle-class Brazilians, for whom the *favelas* represent the taboo and danger: "To many Brazilians, *favelas* are dirty, violent, frightening places. But to many foreigners, they are exciting, interesting, and romantic" (Downie). The newcomers learned about *favelas* primarily through watching films, hence my designation of this kind of impact of films in the global marketplace the "*favela* effect." If the "train effect" marked the changes of time, place, and perception within national contexts of modern discourses on cities, the *favela* effect marks the larger processes by which films are created, circulated, and received in the postmodern context of global capitalism. The *favela* effect does not just result from the global distribution and reception of film; it also describes a complex dynamic in which national cinemas incorporate the styles and genres of other traditions in order to produce these representations of their own culture, which are then distributed as "authentic" to a global audience.

Like the train effect, the *favela* effect has much to tell us about how films about the city are produced, distributed, and received. Because there are many ways to approach these questions, I use *City of God* as an example for thinking about how the study of films about the city can be researched in papers and other scholarly projects. The pairing of this film and the related social phenomenon, the *favela* effect, serves as a springboard to research questions tied to concepts described in the foregoing chapters.

The process of conducting research for a research paper begins with the choice of a topic that, most importantly, should be of interest to the author and fit the scope of the assignment, be it a semester research paper, an MA thesis, a dissertation, an article, or a book manuscript. The scope of the topic should also correlate to the time available to write a paper – three weeks, a semester, an academic year – and the requisite word-count of the paper. Beginning writers often choose a research question or theme that is too broad for the time available and the required length of the paper, and it is important to remember that the topic can be adjusted if this proves to be the case.

Authors then need to decide whether a topic necessitates primary or secondary research. Even though most students do not conduct primary research for a term paper, it is necessary to understand the difference between primary and secondary research in order to recognize both in scholarly literature. *Primary* research constructs a thesis and advances an argument on the basis of research conducted by the writer herself, often in the original location or based on original documents, such as texts, demographics, interviews, surveys, audience analysis, and statistical data. *Secondary* research, on the other hand, consists of researching texts written by others.

Primary research may be qualitative or quantitative. In the present case, *quantitative* research could, for example, refer to representative samples of interviews with inhabitants of the *favela* that are evaluated according to data sets and compared across categories, such as time, class, and race. *Qualitative* social research, then, might entail in-depth interviews of a small number of research subjects, for instance, members of three generations of inhabitants of the City of God, which are then interpreted according to topics or categories that emerge from the interviews. The example at hand illustrates two particular difficulties beyond the regular challenges associated with primary research: the geographical distance, which takes time and money to overcome and may include a language barrier; and the fact that people engaging in illegal activity often do not allow researchers access to information, so researchers must depend on documents by such authorities such as the police, the military, and the city administration.

Secondary research does not encounter such challenges. The challenges of secondary research lie only in finding materials and using the correct key terms and concepts in searching for essays and books that help conceptualize a project and answer research questions adequately. It is more important for secondary research to clarify whether materials are available. Primary and secondary research can be integrated into one project and hence one research paper. For example, a paper on policing in racially or ethnically marked neighborhoods can integrate theoretical models about the role of policing in a modern society, news reports, and statistical data with interviews of residents.

Generally, however, the only primary text in a term paper on film is the *film* itself, which the writer interprets in relation to secondary sources. Indeed, that is the approach used throughout the case studies in this book. Outstanding research papers draw greater implications from a single film or a small number of films or take an aspect within a film (e.g. a recurring motif) and work outward from there. All of the films discussed in this book are available in distribution; however, some films – for example several of the early films mentioned in the Introduction – are available only in archives, which also house and collect documents such as screenplays, shooting diaries, letters, and correspondence between producers and directors. Research on those kinds of documents is also considered primary research and is often undertaken by film scholars, particularly film historians, and is therefore generally beyond the scope of a regular term paper. For a term paper, writers should begin by checking the availability of primary texts (films on DVD or video) and of secondary literature on the chosen topic in order to determine whether the information needed is available. If none or too little is available, the topic can be adjusted or even changed.

So let us now turn to the film *City Of God* and the different ways in which it could serve as the focal point for research papers on cinema and the city. *City Of God* portrays a gang-infested, poor, urban environment called the City of God, which serves as the setting for action-packed, fast-paced, ghetto narrative organized around the film's teenage narrator, Rocket, who becomes a photographer for a Rio de Janeiro newspaper. As a stand-in for the director, Rocket documents the escalation of gang violence and comments in a voice-over on the history of the City of God – which, however, is not a *favela* in the strictest sense of the word, since it was constructed, ironically, by the government in the 1960s as a planned community in an attempt to rid Rio of *favelas*. In any case, the City of God is surrounded by several *favelas* and shares their substandard living conditions. The film references this history by showing the City of God in the 1960s as an obviously planned environment consisting of regular streets lined with small identical houses. These images contrast with the portrayal of the contemporary City of God as a crime-infested maze, in which criminals fight with each other in labyrinthine streets and street kids live in shacks. Paradoxically, a film such as *City of God* is influenced by the US ghetto genre that is now geared towards global distribution and reception. When US, British, and German tourists travel to Rio and insist on knowing the "real Brazil" better than their middle-class Brazilian counterparts, these global tourists import their own visions of the Brazilian *favela* back to Brazil. The claim of authenticity that the film advances and that the tourists evoke and enact, however, is rooted in the US convention of the ghetto film.

To begin thinking about this film one might return to the section entitled "How to read a city?" (p. 15) in the Introduction and consider the film's relationship to the particular city it depicts. Cities are composed of material and symbolic spatial relations which may apply to forces such as the police, the military, and city services. How are their resources distributed across the city and what are their symbolic functions? Does this film reproduce the spatial divisions that create the *favela*? That is, how does the urban structure produce and reproduce social stratification? Does the film criticize the ways in which the represented society is stratified, or does it reproduce the spatial divisions imposed by the dominant society? These kinds of questions lend themselves well to approaching a film that depicts urban space.

Any term paper organized around a film or set of films must also be sure to treat it or them as cinematic products, that is, as documents that use cinematic techniques to convey meaning. The section in the Introduction headed "How to read a film?" (16) explained the analysis of cinematic techniques in the construction of urban space. A discussion of *City of God* in cinematic terms would begin with the ways in which the particular social reality and urban space are staged cinematically. Opening sequences, for instance, often organize the binaries that structure films in spatial and narrative terms. The beginning of *City of God* employs the documentary convention of the voice-over, in which an omniscient narrator compares the past of the ghetto with its present. We encounter the contrast between the inside – the domestic space inhabited by women which will later be taken over by criminals so that there is no space is left uninhabited by criminality – and the outside, which is visually limited and thus reflects the restrictions of the ghetto. The past is shot in warm colors that approximate the cinematography of the time period that is evoked and is edited at a slower pace than the rest of the narrative to imply not only the change of history, but also the change in how time is experienced at different points in history.

In addition to these fundamental approaches, the three sections of this book also offer models for organizing a paper around *City of God*. Section I considers national cinemas and the particular role accorded to representative cities in defining the "imagined community" of the nation. To discuss the representation of a city in terms of the nation, a comparison of films can outline characteristics by contrasting and interpreting them in the context of socio-historical changes, also considering changes in cinematic style in the portrayal of social reality. An interdisciplinary approach helps to investigate the question whether changes on film reflect *social* change or only mirror change in *visual* culture, or both, and what role they play for the "imagined community" of the nation.

Section I also outlines a history of cinematic representation of cities, describing how Weimar cinema defined Berlin as a locus of larger cultural developments, how film noir enlisted Los Angeles to portray postwar society, and how the French New Wave used Paris to outline French social structures. How does *City of God* – or any film about a city – make its urban space represent larger societal structures at its historical moment and pertinent to its geographic location? In *City of God*, for instance, the visual depiction of the city constructs the urban space to show the extreme poverty and violence of the Third World.

And Section I also emphasizes privileged, urban, cinematic spaces, such as the street. Any analysis of a city film can begin with an examination of the primary urban spaces portrayed in the film. How are spaces staged and linked or not linked to other spaces in the city? How are interior and exterior spaces, public and private spaces, defined? How do spaces reflect the social reality of race, class, and gender? An analysis of *City of God* might find that streets increasingly become the sites of violent encounters, and that the film increasingly erases any meaningful distinction between public and private spaces.

Section II emphasizes the role of history for films from Hong Kong, films about war-torn cities, and science-fiction films. Chapter 4 in particular addresses the body politics that created Hong Kong action cinema. Similarly, the slave trade to Brazil as a former colony of Portugal created a hybrid martial arts form, *capoera*, and a hybrid religion, *candomblé*. Do we see any traces of these indigenous hybrid cultural forms? Where are they located in the narrative? Chapter 5 summarizes Stephen Graham's argument that attacks on a city in the context of national war exist on a continuum, from internal urban division, discrimination, policing, to violent attacks on minorities, immigrants, and the urban poor. This thesis could be investigated in regard to another socio-political and historical continuum of violence against minorities – in the case of Brazil, in relation to its colonial past and its history of military dictatorship. Often, noticing a lack or something missing in a film can lead to significant interpretive insights as well. Hence Chapter 6, on utopian and dystopian film, leads us to ask why so few Third World countries bring forth science-fiction films and why *City of God*, like so many films from the Third World, is indebted to a documentary tradition. Such broad considerations can frame more concrete projects, such as an analysis of documentary aspects in select feature films that take place in cities in Third World countries.

The sections "Ghettos and barrios" (pp. 153–75) and "The global city and cities in globalization" (pp. 195–209) provide perhaps the most obvious theoretical frameworks for analyzing *City of God*. For a larger framework, conditions in Brazil could be compared to other so-called Third World countries or Third World conditions embedded within in the industrial West. Taking a cue from the film's theme of criminality in the form of drugs and guns, a research project might question the social reality of the film's representation: How do drugs circulate in a city? What are the possibilities and limitations of movement for those who live in the *favela*? How are the *favelas* in Brazil circumscribed and policed? How do the power differentials of race, gender, ethnicity, and stage in the life-cycle play out in the topography of a city like Rio and its *favelas*? After researching those issues, one could return to the film to ask how similarly or differently they are depicted in *City of God*.

The convention of the ghetto film as outlined in Chapter 7 portrays the ghetto as a space primarily of the young. In this instance an interdisciplinary approach could productively critique the genre. Ghetto films are importantly defined by a claim to authenticity. How does *City of God* invoke authenticity and to what end? What are the conventions of authenticity, particularly in regard to the city, and how have they changed over time? I have discussed other films that employ particular cinematic techniques to claim an authentic representation of a city, for example in the French New Wave. François Truffaut's *400 Blows* (1959) claims to represent Paris in ways that are closely tied to modernity, with slow-paced and uninterrupted tracking shots through the city. In contrast, *City of God* manipulates time and space in obvious ways, making use of such postmodern features as "bullet time," slow motion, and quick jump-cuts, at the same time that locations appear as labels on the image to create a news-style claim to reality. The voice-over narration is the subjective expression of the film's final claim that the story is based on real events. These similarities in authenticity, urban space, and movement, and the differences of style, location, and history call for interpretation.

City of God is not only a transnational ghetto film but, like many other ghetto films and city films in general, is also a male coming-of-age story, and thus it shares elements with such films as *Boyz N the Hood* and *The 400 Blows*. How is male subjectivity created through the city? All three films claim an autobiographical self-reflexivity. Paradoxically, ghetto narratives present the ghetto as a place from which there is no escape, but the only access to insider knowledge an audience can gain – being, by definition, outside – is from those who have left. Remembering Truffaut's self-referentiality, how does self-referentiality function in *City of God*? In *The 400 Blows* Truffaut appears in a cameo in a carnival ride reminiscent of early cinema's zoetrope, while Rocket, the alter ego of *City of God*'s director Merielles, becomes a photographer. Truffaut made an autobiographical claim to the cinematic space of Paris and to its history by portraying his childhood. How does Merielles create an authorial signature in *City of God*?

An academic project can follow the questions raised by the film or it can confront a film's blindspot. The insights of Chapter 8 about the sexualization of urban space can be used to ask why the transnational ghetto film so consistently reinscribes a highly gendered heterosexuality even though – as Chapter 8 shows – a historical link connects homosexuality, the sex trade, and the ghetto. At the same time, the transnational ghetto film traditionally pits the male bond against the heterosexual relationship. These kinds of topics can be analyzed in a comparative study of transnational ghetto films and also compared to the emerging transnational gay and lesbian cinema discussed in the same chapter.

Chapter 9 contextualizes cities in globalization but also takes account of the global exchange of films and the production of films for a global market. Global tourism and migration are also at stake in the *favela* effect. What is the effect of tourism on Brazil? How is Brazil in general, and Rio in particular, sold to tourists? An emphasis on visual culture can lead to an analysis of advertisements, such as posters and short films. How is sexuality used in those kinds of ads? Which cities are shown and which sections of those cities? What are patterns of sexual tourism and how are they related to poverty globally and locally? Similarly, what is the connection between global consumption of drugs and local consumption and distribution?

Films of the same genre also lend themselves to comparison. The transnational ghetto films discussed in Chapter 7 include John Singleton's *Boyz N the Hood* (1991), Gavin Hood's *Tsotsi* (2005), and Pierre Morel's *Banlieue 13* (2004). A comparison with a focus on cinematic techniques, the setting, the *mise-en-scène*, cinematography and editing can describe the cinematic parameters of the genre. How do these films make their stories palatable and consumable globally? Are they postmodern? What do those films share with styles from advertising and games? Answers to these questions contextualize the emerging genre in local and global contexts. A close reading of the *City of God's* soundtrack could trace the relation between local traditions of Afro-Brazilian music and global trends in electronic soundtrack and determine whether the soundtrack is defined by tension, dialogue, or hybridity.

These are the kinds of questions that should emerge from the approach used in this book. Such questions, however, should lead not only to exciting, passionate, smart, well-researched, and well-written research papers, but also to a different kind of *seeing*, applicable both to film and also to the cultural representation of cities, urban space, and social reality in more general ways. Analyzing the techniques of cinematic illusion, including its philosophical and social dimensions, can destroy the naïve pleasure that films often assume for their audience. As a reward, however, it promises an intellectual pleasure that was already embedded in the symbiotic relationship of cities and cinema from the outset.

Further reading

The following are introductory texts for the analysis of film.

David Bordwell and Kristin Thompson (2008) *Film Art: An Introduction*, New York: McGraw-Hill.

Robert Kolker (2006) *Film, Form, and Culture*, Boston, MA: McGraw-Hill.

William H. Phillips (2005) *Film: An Introduction*, Boston, MA: Bedford.

Notes

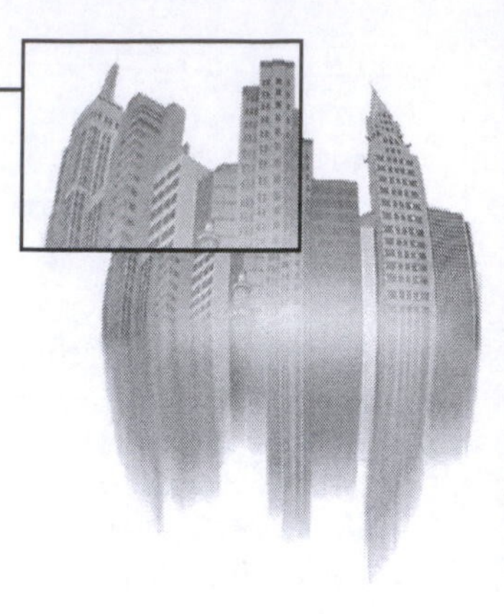

1 This subheading echoes the title of Monaco's book, which is, however, a more extensive and less pragmatic approach to "how to read a film."
2 Most films mentioned in this book are widely distributed and available on DVD or video from major distributors, except for the early short films discussed in the Introduction. This DVD collection provides a good overview of the short films by the directors mentioned in this Introduction and is available for purchase. All titles of international films discussed in the book are given in English. For the additional original titles, see Filmography.
3 In the DVD available in the United States, the brothel is called Madame Gill's Bar. In the German version discussed by Patrice Petro, the name is Frau Greifer. Petro also discusses the censorship that led to the existence of different versions of the film (1990). Parts of the plot mentioned in this section are also not included on the DVD available in the USA, but otherwise all discussions in this book are based on copies of films easily and widely available. With the restoration and re-release of early films, as well as director's cuts, sometimes different versions of films are in circulation. In the future, readers might have access to versions that more accurately reflect the original version of a film discussed in this book. I point out the existence of different versions only in instances in which there are such significant, sometimes factual, differences, as in this one, which could confuse readers.
4 Critics emphasize the book by Raymond Borde and Étienne Chaumeton, *Panorama du film noir américain* (Paris: Editions de Minuit, 1955), translated by Paul Hammond as *A Panorama of American Film Noir, 1941–1953* (San Francisco, CA: City Lights, 2002). However, two other French essays employed the term 'film noir' prior to the publication of their book: Jean-Pierre Chartier, "Les américains aussi font des film noirs," (Americans also make films noir) *Revue du cinèma* 2 (November 1946), and Henri François Ray, "Demonstration par l'absurde: les films noirs," (Demonstration through absurdity: the films noir) *L'ecran français* 157 (June 1948).
5 I thank Susan Hegeman for pointing this out to me.
6 Agnès Varda has a more complicated relationship with the French New Wave than the other directors mentioned here. She is central to this chapter, however, so I have included her in this list and will elaborate on her relationship to the French New Wave later.

7 In French literature on cinema, the following texts discuss the relationship of Paris to the New Wave: "Autour de la nouvelle vague, ou Paris retrouvé," (For the *auteur* of the nouvelle vague, a rediscovery of Paris) in Binh; "La nouvelle vague est arrivé, 1958," (The nouvelle vague has arrived, 1958) in Douchet et Nadeau; and Hillairet *et al.*.

8 The naming of locations in this chapter follows Dirk and Sowa, the only book available that lists locations of 600 films that take place in Paris. It is available only in German.

9 All translations follow the subtitles.

10 I thank Jaimey Fisher for discussing the term with me.

11 I thank Holly Raynard for making this point to me.

12 The script gives this background information about the character:

> Anna Schmidt:
> An Estonian (Czechoslovakian), and therefore officially a Russian citizen, she has been living in Vienna and working as a small part actress under the protection of forged Austrian papers procured for her by Harry Lime, whom she loves. Unlike Martins, she has few illusions about Harry. She has loved him for what he is and not for what she has imagined him to be, and his death leaves her completely indifferent as to her own fate.
>
> (Greene 7)

Bibliography

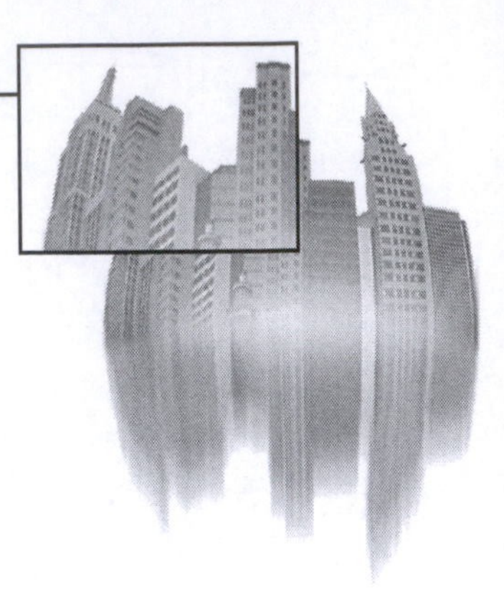

Adler, J.S. (2006) *First in Violence, Deepest in Dirt: Homicide in Chicago, 1875–1920*, Cambridge, MA: Harvard University Press.

"The Advance of Kung Fu," *Asiaweek* (May 9, 1980): 42.

Albrecht, D. (1986) *Designing Dreams: Modern Architecture in the Movies*, New York: Harper & Row.

An, J. (2001) "*The Killer*: Kult Film and Transcultural (Mis)Reading," in E.C.M. Yau (ed.) *At Full Speed: Hong Kong Cinema in a Borderless World*, Minneapolis: University of Minnesota Press, 95–113.

Anderson, B. (1983, 2002) *Imagined Communities*, London: Verso.

Andrew, D. (ed.) (1987) *Breathless: Jean-Luc Godard, Director*, New Brunswick, NJ: Rutgers University Press.

Appadurai, A. (1996) *Modernity at Large: Cultural Dimensions of Globalization*, Minneapolis: University of Minnesota Press.

Armes, R. (1987) *Third World Filmmaking and the West*, Berkeley: University of California Press.

Assmann, A., M. Gomille, and G. Ripple (eds) (2002) *Ruinenbilder* (*Pictures of Ruins*) Munich: Wilhelm Fink Verlag.

Astruc, Alexandre (1948) "The Birth of a New Avant-Garde: La Caméra-Stylo," *L'ecran français* 141; reprinted in (1968) P. Graham (ed.) *The New Wave*, Garden City, NJ: Doubleday, 17–22.

Barber, S. (2003) *Projected Cities: Cinema and Urban Space*, London: Reaktion Books.

Baudrillard, J. (1992) "From *Simulations*," in P. Waugh (ed.) *Postmodernism: A Reader*, London: Arnold, 186–8.

Beck, U. (2000a) "The Cosmopolitan Perspective: On the Sociology of the Second Age of Modernity," *British Journal of Sociology* 51 (1): 79–105.

—— (2000b) *What Is Globalization?* Oxford: Blackwell.

Bell, D., J. Binnie, R. Holliday, R. Longhurst, and R. Peace (2001) *Pleasure Zones: Bodies, Cities, Spaces*, New York: Syracuse University Press.

Bell, D. and G. Valentine (1995) *Mapping Desire*, London: Routledge.

Belton, J. (ed.) (1996) "The Production Code," in *Movies and Mass Culture*, New Brunswick, NJ: Rutgers University Press, 135–49.

Benjamin, W. (1999a) "Paris, Capital of the Ninteenth Century: Exposé of 1939," *The Arcades Project*, Cambridge, MA: Harvard University Press, 14–26.

—— (1999b) "Paris, Capital of the Nineteenth Century: Exposé of 1935," in *The Arcades Project*, Cambridge, MA: Harvard University Press, 3–13.

—— (1999c) "The Return of the Flâneur," in *Selected Writings*, vol. 2, Cambridge, MA: Harvard University Press, 262–7.

—— (2003a) "The Work of Art in the Age of its Technological Reproducibility (Third Version)," in *Selected Writings*, Cambridge, MA: Harvard University Press, vol. 4, 251–83.

—— (2003b) "The Paris of the Second Empire in Baudelaire," *Selected Writings*, vol. 4, Cambridge, MA: Harvard University Press, 3–92.

Binh, N.T. with F. Garbaz (2003) "Autour de la nouvelle vague, ou Paris retrouvé," (For the auteur of the nouvelle vague, a rediscovery of Paris) in *Paris au cinéma: la vie rêvéde la capitale de Méliès à Amélie Poulain*, Paris: Éditions Parigramme, 145–51.

"Blaxploitation Revisited," (2005) *Screening Noir: Journal of Black Film, Television, and New Media Culture* 1(1) (fall/winter).

Bloom, P. (2006) "Beur Cinema and the Politics of Location: French Immigration Politics and the Naming of a Film Movement," in E. Ezra and T. Rowden (eds) *Transnational Cinema: The Film Reader*, London: Routledge, 131–141.

Borde, R. and É. Chaumeton (1955) *Panorama du film noir américain*, Paris: Éditions de Minuit, trans P. Hammond (2002) as *A Panorama of American Film Noir, 1941–1953*, San Francisco, CA: City Lights.

Bordwell, D. (2000) *Planet Hong Kong: Popular Cinema and the Art of Entertainment*, Cambridge, MA: Harvard University Press.

Bordwell, D. and K. Thompson (2008) *Film Art: An Introduction*, New York: McGraw-Hill.

Bottomore, S. (1999) "The Panicking Audience? Early Cinema and the 'Train Effect'," *Historical Journal of Film, Radio and Television* 19 (2), 177–216.

Brodnax, M. (2001) "Man a Machine: The Shift from Soul to Identity in Lang's *Metropolis* and Ruttmann's *Berlin*," in K.S. Calhoon (ed.) *Peripheral Visions: The Hidden Stages of Weimar Cinema*, Detroit, MI: Wayne State University Press, 73–93.

Buck-Morss, S. (1989) *The Dialectics of Seeing: Walter Benjamin and the Arcades Project*, Cambridge, MA: MIT Press.

Bukatman, S. (1997) *Blade Runner*, London: British Film Institute.

Butler, C. (2002) *Postmodernism: A Very Short Introduction*, Oxford: Oxford University Press.

Butler, J. (1993) "Gender Is Burning: Questions of Appropriation and Subversion," in *Bodies that Matter: On the Discursive Limits of 'Sex'*, New York: Routledge, 121–40.

Carter, E., J. Donald, and J. Squires (eds) (1993) *Space and Place: Theories of Identity and Location*, London: Lawrence & Wishart.

Chartier, J.-P. (1946) "Les américains aussi font des film noirs," (Americans also make films noir) *Revue du cinèma* 2 (November), 67–70.

Christie, I. (1994) *The Last Machine: Early Cinema and the Birth of the Modern World*, London: BBC–British Film Institute.

Christopher, N. (1997) *Somewhere in the Night: Film Noir and the American City*, New York: Free Press.

Clarke, D.B. (ed.) (1997a) *The Cinematic City*, London: Routledge.

—— (1997b) "Introduction: Previewing the Cinematic City," in D.B. Clarke (ed.), *The Cinematic City*, London: Routledge, 1–18.

Clarke, D.B. and M. A. Doel (2005) "Engineering Space and Time: Moving Pictures and Motionless Trips," *Journal of Historical Geography* 31, 41–60.

Clover, J. (2004) *The Matrix*. London: British Film Institute.

Cornell, J.C. (1999) "Different Countries, Different Worlds: The Representation of Northern Ireland in Stewart Parker's *Lost Belongings*," in J. MacKillop (ed.) *Contemporary Irish Cinema: From The Quiet Man to Dancing at Lughnasa*, Syracuse, NY: Syracuse University Press, 71–84.

Costabile-Heming, C.A., R.J. Halverson, and K.A. Foell (eds) (2004) *Berlin: The Symphony Continues – Orchestrating Architectural, Social, and Artistic Change in Germany's New Capital*, Berlin: Walter de Gruyter.

Dadameah, S.A. (1972) "Hong Kong's Booming Movie Business," *Penang* (Malaysia) *Star*, August 2: 13.

Dickos, A. (2002) *Street with No Name: A History of the Classic American Film Noir*, Lexington: University Press of Kentucky.

Dimendberg, E. (2004) *Film Noir and the Spaces of Modernity*, Cambridge, MA: Harvard University Press.

Dirk, R. and C. Sowa (2003) *Paris im Film* (Paris on Film), Munich: Belleville.

Doane, M.A. (1991) *Femmes Fatales: Feminism, Film Theory, Psychoanalysis*, New York: Routledge.

Douchet, J. and G. Nadeau (1987) "La nouvelle vague est arrivé, 1958 . . . ," (The nouvelle vague has arrived, 1958) in their *Paris cinéma: Une ville vue par le cinéma de 1895 à nos jours*, Paris: Editions du May, 176–90.

Du Bois, W.E.B. (1996) "The Negro Problem of Philadelphia" (1899), in R.T. LeGates and F. Stout (eds) *The City Reader*, London: Routledge, 119–25.

Downie, A. (2007) "The Pulse of Rio de Janeiro's Slums Luring Foreign Guests," *Christian Science Monitor*, February 6; available: http://www.csmonitor.com/2007/0206/p01s04-woam.html (accessed February 18, 2007).

Dyer, R. (1990) *Now You See It: Studies on Lesbian and Gay Film*, New York: Routledge.

Eisenschmidt, A. and J. Mekinda (n.d.) "Architecture as a Document of Historical Change: Three Examples from Post-War Europe," *Zeithistorische Forschungen/Studies in Contemporary History*; available: http:www.zeithistorische-forschungen.de/site/40208278/default.aspy (accessed September 3, 2007).

Elsaesser, T. (2000a) *Weimar Cinema and After: Germany's Historical Imaginary*, London: Routledge.

—— (2000b) *Metropolis*. London: British Film Institute.

Elsaesser, T. with A. Baker (1990) *Early Cinema: Space – Frame – Narrative*, London: British Film Institute.

Ezra, E. and T. Rowden (eds) (2006) *Transnational Cinema: The Film Reader*, London: Routledge.

Feinstein, H. (1964) "An Interview with Jean-Luc Godard," *Film Quarterly* 17 (3): 8–10.

Fischer, L. (1998) *Sunrise: A Song of Two Humans*, London: British Film Institute.

Fisher, J. (2005) "Wandering in/to the Rubble-Film: Filmic Flânerie and the Exploded Panorama After 1945," *German Quarterly* 78 (4): 461–80.

Flitterman-Lewis, S. (1996) *To Desire Differently: Feminism and the French Cinema*, New York: Columbia University Press.

Forbes, J. (1992) *The Cinema in France: After the New Wave*, London: British Film Institute.

Fore, S. (2001) "Life Imitates Entertainment: Home and Dislocation in the Films of Jackie Chan," in E.C.M. Yau (ed.) *At Full Speed: Hong Kong Cinema in a Borderless World*, Minneapolis: University of Minnesota Press, 115–41.

Frisby, D. (2001) *Cityscapes of Modernity: Critical Explorations*, Cambridge, MA: Polity Press.

Fritzsche, P. (1996) *Reading Berlin 1900*, Cambridge, MA: Harvard University Press.

Garland, D. (2001) *The Culture of Control: Crime and Social Order in Contemporary Society*, Chicago, IL: University of Chicago Press.

Gildea, R.(1997) *France Since 1945*, Oxford: Oxford University Press.

Gleber, A. (1997). "Female Flânerie and the Symphony of the City," in K. von Ankum (ed.) *Women in the Metropolis: Gender and Modernity in Weimar Culture*, Berkeley: University of California Press, 67–88.

—— (1999) *The Art of Taking a Walk: Flânerie, Literature, and Film in Weimar Culture*, Princeton, NJ: Princeton University Press.

Gever, M., J. Greyson, and P. Parmar (eds) (1993) *Queer Looks: Perspectives on Lesbian and Gay Film and Video*, New York: Routledge.

Goebel, R.J. (2001) *Benjamin Heute: Großstadtdiskurs, Postkolonialität und Flanerie zwischen den Kulturen*, Munich: Iudicium.

Graham, S. (2004a) "Introduction: Cities, Warfare, and States of Emergency," in S. Graham (ed.) *Cities, War and Terrorism: Towards an Urban Geopolitics*, Oxford: Blackwell, 1–25.

—— (2004b) "Constructing Urbicide by Bulldozer in the Occupied Territories," in S. Graham (ed.) *Cities, War and Terrorism: Towards an Urban Geopolitics*, Oxford: Blackwell, 192–213.

Grant, B. K. (1995) *Film Genre Reader II*, Austin: University of Texas Press.

Greene, G. (1968, 1988) *The Third Man*, London: Faber & Faber.

Greene, N. (2004) "Representations 1960–2004. Parisian Images and National Transformations," in M. Temple and M. Witt (eds) *The French Cinema Book*, London: British Film Institute, 247–55.

Guerin, F. (2005) *A Culture of Light: Cinema and Technology in 1920s Germany*, Minneapolis, MN: Minnesota University Press.

Gunning, T. (1989) "An Aesthetic of Astonishment," *Art and Text* (spring): 31–45.

—— (2006) "The Birth of Film out of the Spirit of Modernity," in T. Perry (ed) *Masterpieces of Modernist Cinema*, Bloomington: Indiana University Press, 13–40.

Hake, S. (1993) *The Cinema's Third Machine: Writing on Film in Germany: 1907–1933*, Lincoln: University of Nebraska Press.

Hake, S. (1994). "Urban Spectacle in Walter Ruttmann's *Berlin: Symphony of the Big City*," in T.W. Kniesche and S. Brockmann (eds) *Dancing on the Volcano: Essays on the Culture of the Weimar Republic*," Columbia, SC: Camden House, 127–37.

Hall, P. (1998) "The Invention of the Twentieth Century: Berlin 1918–1933," in his *Cities in Civilization*, New York: Pantheon Books, 239–78.

Halle, R. (2002) "German Film, *Aufgehoben*: Ensembles of Transnational Cinema," *New German Critique* 87: 7–46.

Harvey, D. (1989) *The Condition of Postmodernity: An Enquiry into the Origins of Cultural Change*, Malden, MA: Blackwell.

Hayward, Susan (1996, 1999) *Key Concepts in Cinema Studies*, London: Routledge.

Hemphill, Essex (1992) *Where Seed Falls*, in *Ceremonies: Prose and Poetry*, New York: Plume Books.

Hillairet, P., C. Lebrat, and P. Rollet (1985) *Paris vu par le cinéma d'avant-garde 1923–1983*, (Paris seen through Avantgarde Cinema 1923–1983) Paris: Paris Experimental.

Hills, A. (2004) "Continuity and Discontinuity: The Grammar of Urban Military Operations," in S. Graham (ed.) *Cities, War and Terrorism: Towards an Urban Geopolitics*, Malden, MA: Blackwell, 231–46.

Hirsch, F. (1983) *The Dark Side of the Screen: Film Noir*, New York: Da Capo Press.

Hjort, M. and S. Mackenzie (2000) *Cinema & Nation*, London: Routledge.

hooks, b. (1992) "Is Paris Burning?" in b. hooks *Black Looks: Race and Representation*, Cambridge, MA: South End Press, 145–56.

Huyssen, A. (1986) *After the Great Divide: Modernism, Mass Culture, Postmodernism*, Bloomington: Indiana University Press.

—— (2003) *Present Pasts: Urban Palimpsests and the Politics of Memory*, Stanford, CA: Stanford University Press.

Ingram, G. B., A.-M. Bouthillette, and Y. Retter (eds) (1997) *Queers in Space: Communities, Public Places, Sites of Resistance*, Seattle, WA: Bay Press.

Jameson, F. (1998) *The Cultural Turn: Selected Writings on the Postmodern 1983–1998*, London: Verso.

Jarvie, I.C. (1977) *Window on Hong Kong: A Sociological Study of the Hong Kong Film Industry and its Audience*, Hong Kong: Centre for Asian Studies, University of Hong Kong.

Jayamanne, L. (2005) "Let's Miscegenate: Jackie Chan and His African-American Connection," in M. Morris, S.L. Li, and S.C.C. Ching-kiu (eds) *Hong Kong Connections: Transnational Imagination in Action Cinema*, Durham, NC: Duke University Press, 151–62.

Johnston, C. (1998) "*Double Indemnity*," in E.A. Kaplan (ed.) *Women in Film Noir*, London: British Film Institute, 89–98.

Jones, J. (1991) "The New Ghetto Aesthetic," *Wide Angle* 13 (3–4): 32–43.

Jordan, J. (2003) "Collective Memory and Locality in Global Cities," in L. Krause and P. Petro (eds) *Global Cities: Cinema, Architecture, and Urbanism in a Digital Age*, New Brunswick, NJ: Rutgers University Press, 31–48.

Kaes, A. (1996) "Sites of Desire: The Weimar Street Film," in D. Neumann (ed.) *Film Architecture: Set Designs from Metropolis to Blade Runner*, Munich: Prestel, 26–32.

—— (1998) “Leaving Home: Film, Migration, and the Urban Experience,” *New German Critique* 74 (spring–summer): 179–92.

—— (2000) *M*, London: British Film Institute.

—— (2004) “Weimar Cinema: The Predicament of Modernity,” in E. Ezra (ed.) *European Cinema*, Oxford: Oxford University Press, 59–77.

Kaplan, E.A. (ed.) (1998) *Women in Film Noir*, London: British Film Institute.

Katz, E. (1994) *The Film Encyclopedia*, New York: Harper Perennial.

Kitfield, J. (1998) “War in the Urban Jungles,” *Air Force Magazine* 81 (12): 1–8.

Kolker, R. (2006) *Film, Form, and Culture*, Boston, MA: McGraw-Hill.

Konstantarakos, M. (ed.) (2000) *Spaces in European Cinema*, Exeter: Intellect.

Kracauer, S. (1995) *The Mass Ornament: Weimar Essays*, Cambridge, MA: Harvard University Press.

Kraska, P. (2001) *Militarizing the American Criminal Justice System: The Changing Roles of the Armed Forces and the Police*, Boston, MA: Northeastern University Press.

Kreimeier, K. (1996) *The Ufa Story: A History of Germany's Greatest Film Company 1918–1945*, New York: Hill & Wang.

Krutnik, F. (1991) *In a Lonely Street: Film Noir, Genre, Masculinity*, London: Routledge.

Kuhn, A. (ed.) (1999) *Alien Zone II: The Spaces of Science Fiction Cinema*, London: Verso.

Lalanne, J.-M., D. Martinez, A. Abbas, and J. Ngai (eds) (n.d.) *Wong Kar-wai*, Paris: Éditions Dis Voir.

Lamster, M. (ed.) (2000) *Architecture and Film*, New York: Princeton Architectural Press.

Lebeau, V. (2001) *Psychoanalysis and Cinema: The Play of Shadows*, London: Wallflower Press.

Le Berre, C. (2005) *François Truffaut at Work*, London: Phaidon Press.

Leblans, A. (2001) “Inventing Male Wombs: The Fairy-Tale Logic of *Metropolis*,” in K.S. Calhoon (ed.) *Peripheral Visions: The Hidden Stages of Weimar Cinema*, Detroit, MI: Wayne State University Press, 95–119.

Lefebvre, H. (1991, 2005) *The Production of Space*, Malden, MA: Blackwell.

LeGates, R.T. and F. Stout (eds) (1996, 2003) *The City Reader*, London: Routledge.

Lent, J.A. (1990) *The Asian Film Industry*, Austin: University of Texas Press.

Lin, N.-T. (1979) “Some Trends in the Development of the Post-War Hong Kong Cinema,” *Hong Kong Cinema Survey 1946–1968*, Hong Kong: Urban Council, 15–25.

Lo, K.-C. (2005) *Chinese Face/Off: The Transnational Popular Culture of Hong Kong*, Urbana: University of Illinois Press.

Lyotard, J.-F. (1986) “Answering the Question: What is Postmodernism?” in P. Waugh (ed.) *Postmodernism: A Reader*, London: Edward Arnold, 117–25.

Madanipour, A. (1996) “Editors' Introduction,” in R.T. LeGates and F. Stout (eds) *The City Reader*, London: Routledge, 181–2.

—— (1996) “Social Exclusion and Space,” in R.T. LeGates and F. Stout (eds), *The City Reader*, London: Routledge, 182–8.

Marie, M. (1997, 2003) *The French New Wave: An Artistic School*, trans. R. Neupert, Malden, MA: Blackwell Publishing.

Marshall, S. (2006) “Fragile Formen aus Zeit und Raum: Ruinen im Film,” (Fragile Forms of Time and Space Ruins in Film) *Film-dienst* 1: 44–7.

Martinez, G., D. Martinez, and A. Chavez (eds) (1998) *What It Is . . . What It Was! The Black Film Explosion of the '70s in Words and Pictures*, New York: Miramax.

Massey, D.S. and N.A. Denton (eds) (1993) *American Apartheid: Segregation and the Making of the Underclass*, Cambridge, MA: Harvard University Press.

Massod, P.J. (2003) *Black City Cinema: African-American Urban Experiences in Film*, Philadelphia, PA: Temple University Press.

McLoone, M. (1999) "December Bride: A Landscape Peopled Differently," in J. MacKillop (ed.) *Contemporary Irish Cinema: From The Quiet Man to Dancing at Lughnasa*, Syracuse, NY: Syracuse University Press, 40–53.

—— (2000) *Irish Film: The Emergence of a Contemporary Cinema*, London: British Film Institute.

Monaco, J. (2000) *How to Read a Film: Movies, Media, Multimedia*, Oxford: Oxford University Press.

Morley, D. and K. Robins (1995) *Spaces of Identity*, London: Routledge.

Morris, M., S.L. Li, and S.C. Ching-Kiu (eds) (2005) *Hong Kong Connections: Transnational Imagination in Action Cinema*, Durham, NC: Duke University Press.

Mouton, J. (2001) "From Feminine Masquerade to Flâneuse: Agnès Varda's Cléo in the City," *Cinema Journal* 40 (2): 3–16.

Murray, B. (1993) "The Role of the Vamp in Weimar Cinema: An Analysis of Karl Grune's *The Street*," in S.G. Frieden, R.W. McCormick, V.R. Petersen, and L.M. Vogelsang (eds) *Gender and German Cinema: Feminist Interventions*, vol. 2: *German Film History/German History on Film*, 33–41, Providence: Berg.

Naficy, H. (2001) *An Accented Cinema: Exilic and Diasporic Filmmaking*, Princeton, NJ: Princeton University Press.

Neill, W.J.V. and H.-U. Schwedler (eds) (2001) *Urban Planning and Cultural Inclusion: Lessons from Belfast and Berlin*, New York: Palgrave.

Neupert, R. (2002) *A History of the French New Wave*, Madison: University of Wisconsin Press.

North, M. (2004) "World War II: The City in Ruins," in L. Marcus and P. Nicholls (eds) *The Cambridge History of Twentieth-Century English Literature*, Cambridge: Cambridge University Press, 436–52.

Nunn, S. (2001) "Cities, Space and the New World of Law Enforcement Technologies," *Journal of Urban Affairs* 23 (3–4): 259–78.

Ongiri, A.A. (2005a) "Bruce Lee in the Ghetto Connection: Kung Fu Theater and African Americans Reinventing Culture at the Margins," in D. Shipa, L. Nishime, and T.G. Oren (eds), *East Main Street: Asian American Popular Culture*, New York: New York University Press, 249–61.

—— (2005b) "The Shogun of Harlem: Race, Space, and Urbanity in Black Cast Kung Fu Film," *Screening Noir: Journal of Black Film, Television and New Media Culture* 1 (1): 99–113.

Oren, T.G. (2003) "Gobbled Up and Gone – Cultural Preservation and the Global City Marketplace," in L. Krause and P. Petro (eds) *Global Cities: Cinema, Architecture, and Urbanism in a Digital Age*, New Brunswick, NJ: Rutgers University Press, 49–68.

" 'Paris Syndrome' Leaves Tourists in Shock: Japanese Visitors Found to Suffer from Psychiatric Phenomenon," Reuters (October 23, 2006), available: http://www.msnbc.msn.com/id/15391010 (accessed February 18, 2007).

Peters, R. (1996) "Our Soldiers, Their Cities," *Parameters* 26 (1): 1–7.

—— (1997) "The Future of Armored Warfare," *Parameters* 27 (8): 1–9.

Petro, P. (1989) *Joyless Streets: Women and Melodramatic Representation in Weimar Germany*, Princeton, NJ: Princeton University Press.

—— (1990) "Film Censorship and the Female Spectator: *The Joyless Street* (1925)," in E. Rentschler (ed.), *The Films of G.W. Pabst: An Extraterritorial Cinema*, New Brunswick, NJ: Rutgers University Press, 30–40.

Petro, P. and L. Krause (eds) (2003) *Global Cities: Cinema, Architecture, and Urbanism in a Digital Age*, New Brunswick, NJ: Rutgers University Press.

Phillips, William H. (2005) *Film: An Introduction*, Boston, MA: Bedford.

Pieterse, J. (2002) "Globalization, Kitsch and Conflict: Technologies of Work, War, and Politics," *Review of International Political Economy*, 9 (1): 1–36.

Ping-kwan, L. (2000) "Urban Cinema and the Cultural Identity of Hong Kong," in P. Fu and D. Desser (eds) *The Cinema of Hong Kong*, Cambridge: Cambridge University Press, 227–51.

Place, J.A. and L.S. Peterson (1976) "Some Visual Motifs of Film Noir," in B. Nichols (ed.) *Movies and Methods: An Anthology*, Berkeley: University of California Press, 325–38.

Poshek, F., and D. Desser (eds) (2000) *The Cinema of Hong Kong: History, Arts, Identity*. Cambridge: Cambridge University Press.

Prashad, V. (2001) *Everybody Was Kung Fu Fighting: Afro-Asian Connections and the Myth of Cultural Purity*, Boston, MA: Beacon Press.

Ray, H.F. (1948) "Demonstration par l'absurde: les films noirs," (Demonstration through absurdity: the films noir) *L'ecran français* 157 (June).

Riviere, J. (1986) "Womanliness as Masquerade," in V. Burgin, J. Donald, and C. Kaplan (eds) *Formations of Fantasy*, New York: Routledge, 35–44.

Rohmer, É. (1981) "Entretien avec Éric Rohmer," (Discussion with Éric Rohmer) *Cahiers du cinéma* 323–4: 29–39.

Rosenau, W. (1997) "Every Room Is a New Battle: The Lessons of Modern Urban Warfare," *Studies in Conflict and Terrorism* 20 (4): 371–94.

Ross, K. (1995) *Fast Cars, Clean Bodies: Decolonization and the Reordering of French Culture*, Cambridge, MA: MIT Press.

Rykwert, J. (2000) *The Seduction of Place: The History and Future of the City*, New York: Vintage Books.

Sammon, P. (1996) *Future Noir: The Making of Blade Runner*, New York: HarperCollins.

Santner, E.L. (1996) *My Own Private Germany: Daniel Paul Schreber's Secret History of Modernity*, Princeton, NJ: Princeton University Press.

Sassen, S. (1998) *Globalization and its Discontents: Essays on the New Mobility of People and Money*, New York: New Press.

—— (2003) "Reading the City in a Global Digital Age – Between Topographic Representation and Spatialized Power Projects," in L. Krause and P. Petro (eds), *Global*

Cities: Cinema, Architecture, and Urbanism in a Digital Age, New Brunswick, NJ: Rutgers University Press, 15–30.
Schickel, R. (1992) *Double Indemnity*, London: British Film Institute.
Schneider, J. and I. Susser (eds) (2003) *Wounded Cities*, Oxford: Berg.
Schrader, P. (1995) "Notes on Film Noir," in B.K. Grant (ed.) *Film Genre Reader II*, Austin: University of Texas Press, 213–26.
Scott, R. (2005) *Ridley Scott: Interviews*, ed. L.F. Knapp and A.F. Kulas, Jackson: University Press of Mississippi.
Shandley, R.S. (2001) *Rubble Films: German Cinema in the Shadow of the Third Reich*, Philadelphia, PA: Temple University Press.
Shiel, M. and T. Fitzmaurice (eds) (2001) *Cinema and the City: Film and Urban Societies in a Global Context*, Oxford: Blackwell.
—— (2003) *Screening the City*, London: Verso.
Shonfield, K. (2000) *Architecture, Film and the City*, London: Routledge.
Short, J.R. (2006) *Urban Theory: A Critical Assessment*, New York City: Palgrave.
Short, J.R. and Y.-H. Kim (1999) *Globalization and the City*, Harlow, Essex: Longman.
Sikov, E. (1998) *On Sunset Boulevard: The Life and Times of Billy Wilder*, New York: Hyperion.
Silver, A. and E. Ward (eds) (1992) *Film Noir: An Encyclopedic Reference to the American Style*, 3rd edn, Woodstock, NY: Overlook Press.
Silver, A. and J. Ursini (eds) (1996) *Film Noir Reader*, New York: Limelight Editions.
Silver, A. and J. Ursini (2005) *L.A. Noir: The City as Character*, Santa Monica, CA: Santa Monica Press.
Silverman, K. and H. Farocki (1998) *Speaking About Godard*, New York: New York University Press.
Simmel, G. (1903) "The Metropolis and Mental Life", in D. Frisby and M. Featherstone (eds) (1997) *Simmel on Culture*, London: Sage, 174–85.
Smith, J. (2004) "Reading the Red Light District: Literary, Cultural, and Social Discourses on Prostitution in Berlin, 1880–1933," dissertation, Indiana University, Bloomington.
Sobchack, V. (1999) "Cities on the Edge of Time: The Urban Science-Fiction Film," in A. Kuhn (ed.) *Alien Zone II: The Spaces of Science Fiction Cinema*, London: Verso, 123–43.
—— (2004) *Screening Space: The American Science Fiction Film*, New Brunswick, NJ: Rutgers University Press.
Soja, E.W. (2000) *Postmetropolis: Critical Studies of Cities and Regions*, Malden, MA: Blackwell.
Staiger, J. (1999) "Future Noir, Contemporary Noir: Contemporary Representations of Visionary Cities," in A. Kuhn (ed.), *Alien Zone II: The Spaces of Science Fiction Cinema*, London: Verso, 97–122.
Stevenson, D. (2003) *Cities and Urban Cultures*, Maidenhead: Open University Press.
Stokes, L.O. and M. Hoover (1999) *City on Fire: Hong Kong Cinema*, London: Verso.
Strathausen, C. (2003) "Uncanny Spaces: The City in Ruttmann and Vertov," in M. Shiel and T. Fitzmaurice (eds) *Screening the City*, London: Verso, 15–40.

Sun, S. (1982) "A Hong Kong Formula for Hollywood Success," *Asia* (November–December): 38–43.

Truffaut, F. (1954) "Une certaine tendance du cinéma français" ("A Certain Tendency of the French Cinema"), *Cahiers du cinéma* 31 (January): 15–28.

Vincendeau, G. (2000) *Stars and Stardom in French Cinema*, London: Continuum.

Wager, J.B. (1999) *Dangerous Dames: Women and Representation in the Weimar Street and Film Noir*, Athens: Ohio University Press.

Ward, J. (2001a) "Kracauer versus the Weimar Film-City," in K.S. Calhoon (ed.) *Peripheral Visions: The Hidden Stages of Weimar Cinema*, Detroit, MI: Wayne State University Press, 21–37.

—— (2001b) *Weimar Surfaces: Urban Visual Culture in 1920s Germany*, Berkeley: University of California Press.

Warner, K.Q. (2000) *On Location: Cinema and Film in the Anglophone Caribbean*, London: Macmillan.

Warren, R. (2004) "City Streets – The War Zones of Globalization: Democracy and Military Operations on Urban Terrain in the Early Twenty-First Century," in S. Graham (ed.) *Cities, War, and Terrorism: Towards an Urban Geopolitics*, Malden, MA: Blackwell, 214–30.

Weber, M. (1992) *The Protestant Ethic and the Spirit of Capitalism* (1905), London: Routledge.

Wegner, P.E. (2002) *Imaginary Communities: Utopia, the Nation, and the Spatial Histories of Modernity*, Berkeley: University of California Press.

Weihsmann, H. (1997) "The City in Twilight: Charting the Genre of the 'City Film' 1900–1930," in F. Penz and M. Thomas (eds) *Cinema & Architecture: Méliès, Mallet-Stevens, Multimedia*, London: British Film Institute, 8–27.

Willett, J. (1978) *Art and Politics in the Weimar Period: The New Sobriety 1917–1933*, New York: Da Capo Press.

Williams, R. (1973) *The Country and the City*, New York: Oxford University Press.

Wilson, E. (1999) *French Cinema Since 1950: Personal Histories*, New York: Rowman & Littlefield.

Wilson, W.J. (1996) "From Institutional to Jobless Ghettos," in R.T. LeGates and F. Stout (eds) *The City Reader*, London: Routledge, 126–35.

Yau, E.C.M. (ed) (2001) *At Full Speed: Hong Kong Cinema in a Borderless World*, Minneapolis: University of Minnesota Press.

Young, J. (1999) *The Exclusive Society*, London: Sage.

Filmography

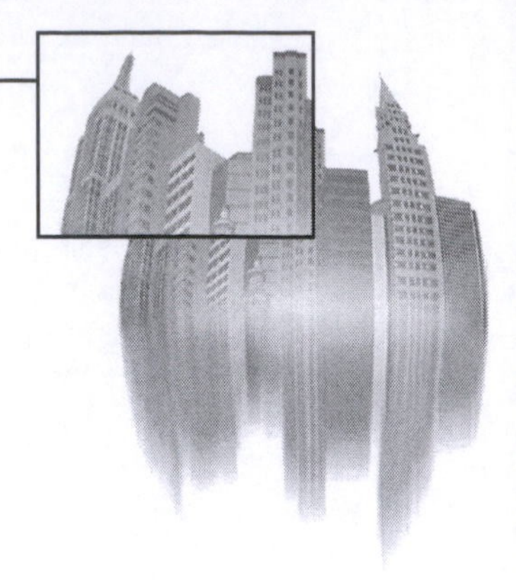

Chantal Akerman. *From the Other Side* (*De l'autre côté*, 2002)
Fatih Akın. *Short Sharp Shock* (*kurz undschmerzlos*, 1998)
—— *In July* (Im Juli, 2000)
—— *Solino* (2002)
—— *Head-On* (Gegen die Wand, 2004)
Robert Aldrich. *Kiss Me Deadly* (1955)
R'anan Alexandrowicz. *James' Journey to Jerusalem* (*Massa'ot James Be'eretz Hakodesh*, 2003)
Thomas Arslan. *Brothers and Sisters* (*Kardesler – Geschwister*, 1997)
E. Kutluĝ Ataman. *Lola and Billy the Kid* (*Lola + Bilidikid*, 1999)
Alain Berliner. *Ma vie en rose* (1997)
Ursula Biemann. *Performing the Border* (1999)
—— *Remote Sensing* (2001)
Stephanie Black. *Life and Debt* (2001)
Craig Brewer. *Hustle & Flow* (2005)
Peter Brooks. *Moderato Cantabile* (1960)
Sue Brooks. *Japanese Story* (2003)
Linda Goode Bryant and Laura Poitras. *Flag Wars* (2003)
Lois Buñuel. *Los Olvidados* (1950)
Charles Burnett. *Killer of Sheep* (1977)
Tim Burton. *Batman* (1989)
Marcel Camus. *Black Orpheus* (1959)
Marc Caro and Jean-Pierre Jeunet. *Delicatessen* (1991)
—— *The City of Lost Children* (*Le cité des enfants perdus*, 1995)
Claude Chabrol. *The Cousins* (*Les cousins*, 1959)
—— *The Good Girls* (*Les bonnes femmes*, 1960)
—— *Handsome Serge* (*Le beau Serge*, 1958)
Gurinder Chadha. *Bend It Like Beckham* (2002)
Fruit Chan. *Made in Hong Kong* (*Xianggang zhizao*, 1998)
Jackie Chan. *Project A* (*'A' gai waak*, 1983)
Madhi Charel. *Tea in the Harem* (1985)

René Clément. *Is Paris Burning?* (*Paris brûle-t-il?*, 1966)
Robert Clouse. *Enter The Dragon* (1973)
Larry Cohen. *Black Caesar* (1973)
Sofia Coppola. *Lost in Translation* (2003)
Robert Culp. *Hickey & Boggs* (1972)
Michael Curtiz. *Casablanca* (1942)
Jules Dassin. *Thieves' Highway* (1949)
Tom Dey. *Shanghai Noon* (2000)
Stanley Donen and Gene Kelly. *Singing in the Rain* (1952)
Richard Donner. *Superman* (1978)
Jean Douchet, Jean Rouch, Jean-Daniel Pollet, Eric Rohmer, Jean-Luc Godard, and Claude Chabrol. *Six in Paris* (*Paris vu par . . .*, 1965)
Ziad Doueiri. *West Beyrouth* (Lebanon, 1998)
Karim Dridi. *Bye-Bye* (1995)
Edison. *What Happened on 23rd Street, New York City* (1901)
Sergei M. Eisenstein. *Battleship Potemkin* (*Bronenosets Potyomkin*, 1925)
William Friedkin. *Cruising* (1980)
Allen Fong. *Father and Son* (*Fu zi qing*, 1981)
John Ford. *Stagecoach* (1939)
Stephen Frears. *Dirty Pretty Things* (2002)
Louis J. Gasnier. *Reefer Madness* (1936)
Karl Gass. *Look at This City!* (*Schaut auf diese Stadt*, 1962)
Haile Gerima. *Bush Mama* (1976)
Bahman Ghobadi. *Turtles Can Fly* (*Lakposhtha hâm parvaz mikonand*, 2004)
Jean-Luc Godard. *Breathless* (*À bout de souffle*, 1960)
—— *Charlotte and Her Boyfriend* (*Charlotte et son Jules*, 1960)
—— *The Little Soldier* (*Le petit soldat*, 1960)
—— *A Woman Is a Woman* (*Une femme est une femme*, 1961)
—— *Contempt* (*Le mépris*, 1963)
—— *Alphaville* (*Alphaville, une étrange aventure de Lemmy Caution*, 1965)
—— *Pierrot le fou* (1965)
—— *Masculine/Feminine: In 15 Acts* (*Masculin/féminin: 15 faits précis*, 1966)
—— *Two or Three Things I Know about Her......* (*Deux ou trois choses que je sais d'elle*, 1966)
D.W. Griffith. *The Musketeers of Pig Alley* (1912)
Tsui Hark. *Dangerous Encounter of the First Kind* (*Diyi leixing weixian*, 1980)
Howard Hawks. *The Big Sleep* (1946)
Perry Henzell. *The Harder They Come* (1972)
Cecil Hepworth. *Explosion of a Motor Car* (1900)
—— *How It Feels to be Run Over* (1900)
Jack Hill. *Coffy* (1973)
Gavin Hood. *Tsotsi* (2005)
Jan Hrebejk. *Up and Down* (*Horem pádem*, 2004)
Ann Hui. *The Secret* (*Feng jie*, 1979)

Ann Hui. *The Boat People* (*Touben nuhai*, 1982)
—— *Summer Snow* (*Nuren shishi*, 1995)
John Huston. *The Maltese Falcon* (1941)
Boris Ingster. *The Stranger on the Third Floor* (1940)
Etang Inyang. *Badass Supermama* (1996)
Isaac Julien. *Looking for Langston* (1988)
Wong Kar-wei. *Chungking Express* (*Conqing senlin*, 1994)
—— *Happy Together* (*Cheun gwong tsa sit*, 1997)
Mathieu Kassovitz. *Café au Lait* (*Mélisse*, 1993)
—— *Hate* (*La Haine*, 1995)
Cédric Klapisch. *Euro Pudding* (*L'auberge espangnol*, 2002)
Gerhard Klein. *A Berlin Romance* (*Eine Berliner Romanze*, 1956).
—— *Berlin – Schönhauser Corner* (*Berlin – Ecke Schönhauser*, 1957)
Stanley Kwan. *Rouge* (*Yin ji kan*, 1987).
—— *Hold Me Tight* (*Yue Kuaile yue duoluo*, 1998)
Ernst Laemmle. *The Devil's Reporter: In the Fog of the City* (1929)
Ringo Lam. *City on Fire* (*Lung fu fong wau*, 1987)
—— *Prison on Fire* 1 (1987)
—— *School on Fire* (1988)
—— *Prison on Fire* 2 (1991)
Gerhard Lamprecht. *Somewhere in Berlin* (*Irgendwo in Berlin*, 1946)
Fritz Lang. *Metropolis* (1927)
—— *M* (1931)
—— *Hangmen Also Die!* (1943)
—— *Ministry of Fear* (1944)
—— *Scarlet Street* (1945)
—— *The Big Heat* (1953)
Bruce Lee. *Way of the Dragon* (1972)
Malcolm D. Lee. *Undercover Brother* (2002)
Mitchell Leisen. *Midnight* (1939)
Dani Levy. *I Was on Mars* (1992)
Jennie Livingston. *Paris Is Burning* (1990)
Wei Lo. *Fist of Fury* (*Jing wu men*, 1972)
Ernst Lubitsch. *To Be or Not to Be* (1942)
—— *Bluebeard's Eighth Wife* (1938)
Louis and Auguste Lumière. *Launching of a Boat* (1890)
—— *The Arrival of a Train at La Ciotat Station* (*L'arrivée d'un train en gare de la Ciotat*, 1895)
—— *Exiting The Factory* (*La Sortie des usines Lumière*, 1895)
Kurt Maetzig. *The Story of a Young Couple* (*Roman einer jungen Ehe*, 1952)
Luigi Maggi. *The Count of Montecristo* (1908)
Mohsen Makhmalbaf. *Kandahar* (*Safar e Ghandehar*, 2001)
Anthony Mann. *T-Men* (1947)
Michael Mann. *Heat* (1995)

Joseph P. Mawra. *Chained Girls* (1965)
Joe May. *Asphalt* (1929)
—— *Music in the Air* (1934)
Fernando Meirelles. *City of God* (2002)
Jean-Pierre Melville. *The Forgiven Sinner* (*Léon Morin, prêtre*, 1961)
—— *Doulos: The Finger Man* (*Le Doulos*, 1963).
William Cameron Menzies. *Things to Come* (1936)
Benny Chan Muk-sing and Jackie Chan. *Who Am I?* (1998)
Ángel Muñiz. *Nueba Yol* (1995)
F.W. Murnau. *The Last Laugh* (*Der letzle Mann*, 1924)
Mira Nair. *Salaam Bombay!* (1988)
Takehiro Nakajima. *Okoge* (1992)
Damien O'Donnell. *East Is East* (1999)
Richard Oswald. *Different from the Others* (*Anders als die Anderen*, 1919)
Ali Özgentürk. *The Horse* (*At*, 1982)
G.W. Pabst. *Joyless Street* (*Die freudlose Gasse*,1925)
Euzhan Palcy. *Sugarcane Alley* (*Rue cases nègres*, 1983)
Brian De Palma. *Dressed To Kill* (1980)
—— *Scarface* (1983)
Gordon Parks. *Shaft* (1971)
Pier Paolo Pasolini. *Accattone* (1961)
—— *Mamma Roma* (1962)
Robert William Paul. *The Last Days of Pompeii* (1897)
—— *Come Along, Do!* (1898)
—— *The Haunted Curiosity Shop* (1901)
——*The ? Motorist* (1906)
—— *A Tour Through Spain And Portugal* (n.d.)
Raoul Peck. *Profit And Nothing But! Or Impolite Thoughts on the Class Struggle* (2001)
Mario Van Peebles. *New Jack City* (1991)
—— *Sweet Sweetback's Baadasssss Song* (1971)
Arthur Penn. *Bonnie and Clyde* (1967)
Gillo Pontecorvo. *Battle of Algiers* (*La Battaglia di Algeri*, 1966)
Edwin S. Porter. *Kansas Saloon Smashers* (1901)
—— *Terrrible Teddy, The Grizzly King* (1901)
—— *The Great Train Robbery* (1903)
Lordes Portilla. *Senorita Extraviada* (2001)
Alexis Proyas. *Dark City* (1998)
Sam Rami. *Spiderman* (2002)
Brett Ratner. *Rush Hour* (1998)
—— *Rush Hour 2* (2001)
Carol Reed. *The Third Man* (1949)
Matty Rich. *Straight out of Brooklyn* (1991)
Leni Riefenstahl. *Triumph of the Will* (*Triumph des Willens*, 1935)
Rintaro. *Osamu Tezuka's Metropolis* (*Metoroporisu*, 2001)

Martin Ritt. *The Spy Who Came in from The Cold* (1965)
Jacques Rivette. *Paris Belongs to Us* (*Paris nous appartient*, 1959)
Mark Robson. *The Harder They Fall* (1956)
George A. Romero. *Dawn of the Dead* (1978)
Roberto Rossillini. *Germany Year Zero* (*Germania Anno Zero*, 1948)
Jean Rouch and Edgar Morin. *Chronicle of a Summer* (*Chronique d'un été*, 1961)
Russell Rouse. *D.O.A.* (1950)
Josef Rusnak. *The Thirteenth Floor* (1999)
Walter (also known as Walther) Ruttmann. *Berlin: Symphony of a Great City* (*Berlin: Die Symphonie der Großstadt*, 1927)
Leontine Sagan. *Girls in Uniform* (Mädchen in Uniform, 1931)
John Schlesinger. *Midnight Cowboy* (1969)
Hans-Christian Schmid. *Lights* (Lichter, 2003)
Martin Scorsese. *Who's That Knocking on my Door?* (1968)
Ridley Scott. *Blade Runner* (1982)
Jim Sheridan. *The Boxer* (1997)
—— *Get Rich or Die Tryin'* (2005)
John Singleton. *Boyz N the Hood* (1991)
Robert Siodmark, Edgar Ulmer, and Billy Wilder. *People on Sunday* (*Menschen am Sonntag*, 1929)
—— *Phantom Lady* (1944)
—— *Criss Cross* (1949)
Jack Snyder. *Dawn of the Dead* (2004)
Wolfgang Staudte. *The Murderers Are among Us* (*Die Mörder sind unter uns*, 1946)
Robert A. Stemmle. *The Ballad of Berlin* (*Berliner Ballade*, 1948)
Josef von Sternberg. *The Blue Angel* (*Der blaue Engel*, 1930)
A. Edward Sutherland. *Champagne Waltz* (1937)
Mak Tai-wai. *The Wicked City* (*Yaoshou Dushi*, 1992)
Quentin Tarantino. *Jackie Brown* (1997)
Lars van Trier. *Europa* (*Zentropa*, 1991)
François Truffaut. *The 400 Blows* (*Les quatre cents coups*, 1959)
—— *Shoot the Piano Player* (*Tirez sure le pianiste*, 1960)
—— *Antoine and Colette* (*Antoine et Colette*, 1962)
—— *Jules and Jim* (*Jules et Jim*, 1962)
—— *Stolen Kisses* (*Baisers volés*, 1968)
—— *Bed and Board* (*Domicile conjugal*, 1970)
—— *Love on the Run* (*L'amour en fuite*, 1978)
Stanley Tong. *Rumble in the Bronx* (1995)
Monika Treut. *Virgin Machine* (*Die Jungfrauenmaschine*, 1988)
—— *My Father Is Coming* (1991)
Eric Tsang. *Aces Go Places* (*Zuijia Paidang*, 1982)
Hdeng Tsu. *Rumble in Hong Kong* (*Nui ging chaat*, 1974)
Agnès Varda. *Cléo from 5 to 7* (*Cléo de 5 à 7*, 1962)
Andy Wachowski and Larry Wachowski. *The Matrix* (1999)

Raoul Walsh. *White Heat* (1949)
Li Min Wei. *Rouge* (1923)
Peter Weir. *The Truman Show* (1998)
James H. White. *Black Diamond Express* (1896)
Robert Wiene. *The Cabinet of Dr Caligari* (*Das Kabinett des Dr Caligari*, 1919)
Billy Wilder. *The Major and The Minor* (1942)
—— *Double Indemnity* (1944)
—— *Sunset Boulevard* (1950)
—— *The Seven-Year Itch* (1955)
Todd Williams. *Friendly Fire: Making An Urban Legend* (2003)
Terence H. Winkless. *The Berlin Conspiracy* (1992)
Doris Wishman. *Bad Girls Go to Hell* (1965)
John Woo. *A Better Tomorrow* (*Ying hung boon sik*,1986)
—— *The Killer* (*Dip hyut shueng hung*, 1989)
—— *Hard-Boiled* (*Laat sau sen taan*, 1992)
Woo-ping Yuen. *Drunken Master* (*Jui Kuen*,1978)
Yu lik-wai. *Love Will Tear Us Apart* (*Tian shang ren jian*, 1999)

Index